TI-83 GRAPHING CALCULATOR GUIDE

for

David S. Moore's

THE BASIC PRACTICE OF STATISTICS

Second Edition

David K. Neal
Western Kentucky University

W. H. Freeman and Company
New York

ISBN 0-7167-3614-4

Printed in the United States of America

Third Printing, 2002

Contents

CHAPTER 12   Nonparametric Tests

Preface

The study of statistics has become commonplace in a variety of disciplines, and the practice of statistics is no longer limited to specially trained statisticians. There are now practitioners in fields such as biology, economics, psychology, sociology, and nursing. Whether a professional statistician or not, all practitioners must rely on the proper use of statistical methods for their work. Few practitioners are inclined to perform the long, tedious calculations that are often necessary in statistical inference. Fortunately, there are now software packages and calculators that can perform many of these calculations in an instant, thus freeing the user to spend valuable time on methods and conclusions rather than on computation.

With its built-in statistical features, the TI-83 Graphing Calculator has dramatically improved the teaching of statistics. Students and teachers can now have instant access to many statistical procedures at the touch of a few buttons.

This manual serves as a companion to *The Basic Practice of Statistics* (Second Edition) by David S. Moore. Problems from the text are worked using either built-in TI-83 features or programs that can be downloaded into the TI-83. The remarkable capabilities of the TI-83 are demonstrated throughout. It is hoped that all students, teachers, and practitioners of statistics discover and make use of these features. Hopefully this manual will be helpful to those who do.

Programs

All codes and instructions for the programs are provided in the manual; they can be downloaded in ASCII format at

http://www.wku.edu/~neal/bps/ti83.html

It is recommended that instructors download these programs to computer and then send them to their calculator via the TI-83 Graph Link cable. Instructors can then send them to students' calculators via direct link.

Acknowledgments

I would like to thank the editorial staff at W. H. Freeman and Company for providing the support and the opportunity to write this manual. I especially offer gratitude to Professor Moore for once again providing educators and students with an excellent text for learning statistics.

David K. Neal
Department of Mathematics
Western Kentucky University
Bowling Green, KY 42101

email: nealdk@wku.edu
homepage: http://www.wku.edu/~neal/neal.html

CHAPTER
1

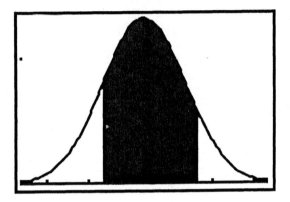

Examining Distributions

We begin by using the TI-83 to store and view data sets. In this chapter, we first see how to make histograms and time plots. Then we learn how to compute basic statistics, such as the mean, variance, standard deviation, median, and quartiles, and how to view data further with boxplots. Last, we study the normal distributions and provide a program to compute normal probabilities and graph a normal curve.

1.1   Displaying Distributions with Graphs

We start by using the TI-83 to help us visualize data sets. In this section, we will use the **STAT** menu to store data sets in lists, and the **STAT PLOT** menu to create histograms and time plots.

1.5 Automobile fuel economy. Make a histogram of the highway gas mileage for the following set of 1998 model midsize cars.

Model	MPG	Model	MPG
Acura 3.5RL	25	Lexus GS300	23
Audi A6 Quattro	26	Lexus LS400	25
Buick Century	29	Lincoln Mark VIII	26
Cadillac Catera	24	Mazda 626	33
Cadillac Eldorado	26	Mercedes-Benz E320	29
Chevrolet Lumina	29	Mercedes-Benz E420	26
Chrysler Cirrus	30	Mitsubishi Diamante	24
Dodge Stratus	28	Nissan Maxima	28
Ford Taurus	28	Oldsmobile Aurora	26
Honda Accord	29	Rolls-Royce Silver Spur	16
Hyundai Sonata	27	Saab 900S	25
Infiniti I30	28	Toyota Camry	30
Infiniti Q45	23	Volvo S70	25

Solution: We must first enter the data into the TI-83.

Entering Data into a List

Step 1:   Press **STAT**, then press **1** to call up the **STAT EDIT** screen.

Step 2:   In order to clear any data that might be in the lists, press the up arrow to highlight **L1**, press **CLEAR**, then press **ENTER**. If desired, highlight **L2**, press **CLEAR**, and press **ENTER**. Then move the cursor back under list **L1**.

 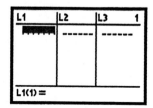

Step 3:   To enter the data into list **L1**, type **25**, press **ENTER**; type **26**, press **ENTER**; continue until the entire data set is entered into **L1**.

L1	L2	L3	1
25	------	------	
26			
29			
24			
26			
29			
30			
L1(7)=30			

Restriction: At most 999 measurements can be entered into a list.

Sorting a List

If desired, you can sort the data into increasing order. Press **STAT**, press **2** to obtain the command **SortA(** on the Home screen. Then type **2nd L1)** and press **ENTER** to execute the command **SortA(L1)**. Now press **STAT**, press **1**, and observe that the data are now in increasing order. Scroll to the bottom of **L1** to see the largest value. By sorting, it is easier to observe the range of the data.

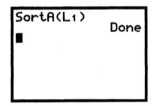

L1	L2	L3	2
16	▬▬▬▬	------	
23			
23			
24			
24			
25			
25			
L2(1)=			

L1	L2	L3	1
29			
29			
29			
30			
30			
33			

L1(26) =33			

Histograms

Now let us make a histogram of our data set where the x axis ranges from 15 to 35 on a scale of 2. After entering the data into a list, we must adjust the **WINDOW** and **STAT PLOT** settings.

Step 1:   Press **WINDOW** and enter the settings as shown. Note that **Ymin** should always be **0** for a histogram, but that the **Ymax** may need to be adjusted in order to see the top of each bar.

```
WINDOW
Xmin=15
Xmax=35
Xscl=2
Ymin=0
Ymax=10
Yscl=1
Xres=1■
```

Note: Since **Xscl** is set at 2, the bars will have width 2. Thus, we set **Xmax** to at least 2 more than the largest measurement. This setting allows us to see the last bar without it being cut off.

Step 2: Press **STAT PLOT (2nd Y=)** and press **1** to get to the settings screen for **Plot1**. Highlight **On** and press **ENTER**. Scroll down to **Type**, then scroll right to highlight the third type (histogram), and press **ENTER**. Set **Xlist** to **L1** and **Freq** to **1** (since we count each data point one time).

Step 3: Before graphing, return to the **STAT PLOT** screen and, if necessary, turn off **PLOT2** and **PLOT3**. Also press **Y=** and either clear or de-select any functions to prevent them from being graphed.

Step 4: Now press **GRAPH**. Then press **TRACE** and move the right arrow cursor along the histogram to see the range of each bar and how many measurements lie in that range.

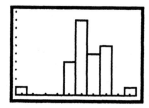 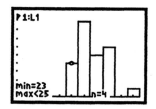

Note: By adjusting **Xscl** in the **WINDOW**, we can alter the shape of the histogram. The graph to the right uses an **Xscl** of 1.

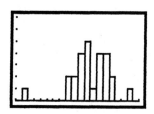

1.80 Guinea pig survival times. Make a histogram of the survival times (in days) of guinea pigs injected with tubercle bacilli in a medical experiment.

43	45	53	56	56	57	58	66	67	73
74	79	80	80	81	81	81	82	83	83
84	88	89	91	91	92	92	97	99	99
100	100	101	102	102	102	103	104	107	108
109	113	114	118	121	123	126	128	137	138
139	144	145	147	156	162	174	178	179	184
191	198	211	214	243	249	329	380	403	511
522	598								

Solution: So as not to lose our previous data in **L1**, we will enter the data from this exercise into list **L2**. Clear list **L2** and enter the data into this list. Although not necessary in this case, we would sort the data by entering the command **SortA(L2)**.

We will graph using a range of 40 to 620 on a scale of 20. After adjusting the **WINDOW**, set **Xlist** to **L2** in the **STAT PLOT** screen. Then press **GRAPH**.

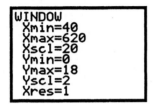  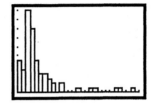

Frequency Charts

Often a data set is given in a *frequency chart*, which lists the number of times each measurement occurs rather than listing each measurement numerous times.

Number of children per household. Make a histogram for the following counts of the number of children living in a household.

Number of children	0	1	2	3	4	5	6
Number of households	60	42	86	59	22	4	2

Solution: To enter the data and not lose our previous data, we will use lists **L3** and **L4**. Clear any data from lists **L3** and **L4**, then enter the counts (children) into **L3** and the frequencies (households) into **L4**.

In the **WINDOW**, we must set the ranges so that we see the entire histogram. We set **Xmin** to 0 and **Xmax** to 7, on a scale of 1. We set **Ymax** to 90 with a **Yscl** of 10. In the **STAT PLOT** screen, we set **XList** to L3 but then set **Freq** to L4. After adjusting the settings, press **GRAPH** and then **TRACE**.

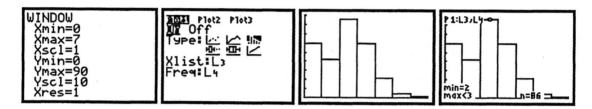

Storing a List in Another List

We may wish to use lists **L1** and **L2** for other data sets but not wish to lose the current data in those lists. We can always move the data from these lists into other lists. To move the data from lists **L1** and **L2** into lists **L5** and **L6**, enter the commands **L1 ➜ L5** and **L2 ➜ L6**. Observe that the original **L1** and **L2** data sets are now in **L5** and **L6**.

Time Plots

We next discuss how to view data observations made over a period of time. Rather than using a histogram, we shall now use a *time plot*.

1.10 Yields of money market funds. Make a time plot of the average annual interest rates (in percent) paid by all taxable money market funds since 1973.

Year	Rate	Year	Rate	Year	Rate	Year	Rate
1973	7.60	1979	10.92	1985	7.77	1991	5.70
1974	10.79	1980	12.88	1986	6.30	1992	3.31
1975	6.39	1981	17.16	1987	6.17	1993	2.62
1976	5.11	1982	12.55	1988	7.09	1994	3.65
1977	4.92	1983	8.69	1989	8.85	1995	5.37
1978	7.25	1984	10.21	1990	7.81	1996	4.80

Solution: We will enter the years into list **L1** and the rates into **L2**. To do so, first follow Steps 1–3 in the solution to Exercise 1.5. Then adjust the **WINDOW** settings so that **X** ranges between the years in listed in **L1** and **Y** ranges appropriately to see the rates in **L2**.

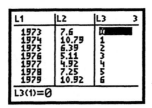 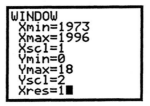

Next adjust the **STAT PLOT** settings. After turning on **Plot1**, press the down arrow to move to **Type**, then scroll right to highlight the second type (time plot), and press **ENTER**. Enter **L1** for **Xlist** and **L2** for **Ylist**. Before graphing, remember to turn off the other plots and either clear or deselect other functions in the **Y=** screen. Then press **GRAPH** and **TRACE**.

 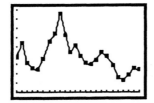 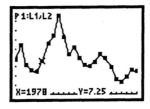

We can now observe the overall fluctuations in average money market rates.

1.22 (b) The Influenza Epidemic of 1918. Make time plots of the weekly new cases (divided by 10) and the number of deaths in San Francisco during the worldwide influenza outbreak of 1918–19.

Date	Oct 5	Oct 12	Oct 19	Oct 26	Nov 2	Nov 9	Nov 16	Nov 23	Nov 30
Cases	36	531	4233	8682	7164	2229	600	164	57
Deaths	0	0	130	552	738	414	198	90	56

Date	Dec 7	Dec 14	Dec 21	Dec 28	Jan 4	Jan 11	Jan 18	Jan 25
Cases	722	1517	1828	1539	2416	3148	3465	1440
Deaths	50	71	137	178	194	290	310	149

Solution: We will denote the weeks as the values 1–17 in list **L1**. Then we enter the numbers of cases divided by 10 into list **L2** and the numbers of deaths into list **L3**. Adjust the **WINDOW** settings as shown and the **Plot1** settings as in the previous exercise. Press **GRAPH** to see the time plot of the numbers of cases (divided by 10).

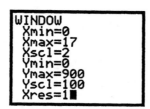 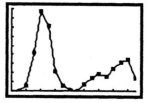

Next, turn off **Plot1** and adjust the **Plot2** settings to graph **L1** versus **L3**. To see both on the same screen, turn on both **Plot1** and **Plot2** and regraph.

 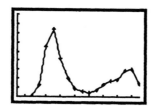 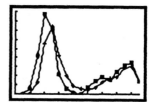

1.2 Describing Distributions with Numbers

In this section, we compute the various statistics of a data set including the mean, variance, standard deviation, median, and quartiles. We also use boxplots to view these statistics.

Basic Statistics

1.41 The density of the earth. Find the mean, variance, standard deviation, median, first quartile, and third quartile of Cavendish's data on the density of the earth. Create a boxplot to view the spread.

5.50	5.61	4.88	5.07	5.26	5.55	5.36	5.29	5.58	5.65
5.57	5.53	5.62	5.29	5.44	5.34	5.79	5.10	5.27	5.39
5.42	5.47	5.63	5.34	5.46	5.30	5.75	5.68	5.85	

Solution: We first enter the data into a list in the **STAT EDIT** screen. For this problem, we shall enter the data into list **L4**. To do so, follow the steps outlined in the previous section under **Entering Data into a List**.

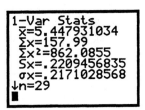

Now press **STAT**, press the right arrow to display the **CALC** screen, and press **1** to bring the command **1-Var Stats** to the Home screen. Press **L4** (**2nd 4**) to obtain the command **1-Var Stats L4**. Press **ENTER**.

Note: If the data had been entered into a different list, say list **L2**, then we would use the command **1-Var Stats L2**.

After completing these steps, we receive a display of the desired statistics:

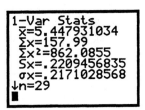

Since the data are a *sample* of measurements, the value of $\bar{x}$ is the sample mean of the population, though it can be considered to be the true mean μ of just this data set. Two standard deviation values are given. The first, **Sx**, is the sample deviation that is to be used if considering these data to be a sample from a larger population. The second, σx, is the true standard deviation of just this set of measurements.

Press the down arrow to scroll down and view more statistics:

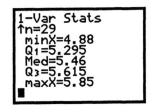

The median is 5.46. That is, 5.46 is the "middle" measurement; around half of these density measurements should be 5.46 or below.

The first quartile is **Q1** = 5.295 and the third quartile is **Q3** = 5.615. Thus, around 1/4 of the measurements are 5.295 or below, while around 3/4 of the measurements are 5.615 or below.

Accessing the Statistics

After computing the statistics, the calculator stores their values in the **VARS Statistics** screen. Whenever needed, we can retrieve the values from this screen.

One deviation from average. For the data of Exercise 1.41, recall the value of $\bar{x}$ and compute the interval $(\bar{x} - Sx, \bar{x} + Sx)$. Then compute the sample variance.

Solution: Press **CLEAR** (or **2nd Quit**) to return to the Home screen. To recall $\bar{x}$, press **VARS**, press **5** for **Statistics**, press **2** for $\bar{x}$, then press **ENTER**.

```
x̄
          5.447931034
x̄-Sx
          5.226985351
x̄+Sx
          5.668876718
```

To compute $(\bar{x} - Sx, \bar{x} + Sx)$, press **VARS**, press **5**, press **2**, press **−**, press **VARS**, press **5**, press **3**, press **ENTER**. We have now entered the command $\bar{x} - Sx$. Now press **2nd ENTER** to recall the previous command, edit it to $\bar{x} + Sx$, and press **ENTER**.

We see that measurements from 5.227 to 5.669 are within one standard deviation of average.

Sample Variance

To find the sample variance, first access **Sx** as before, then press x^2 to obtain the command Sx^2. Then press **ENTER**.

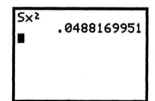

Boxplots

We will continue working with Cavendish's data to make a boxplot. We first set the **WINDOW,** as with a histogram, to the appropriate range of the measurements (although the boxplot ignores the **Y** range).

In the **STAT PLOT** screen, first turn on **PLOT1** (and turn off all other plots and functions). Press the down arrow to get to **Type,** then scroll the right arrow to highlight the boxplot (fifth type), and press **ENTER.** Set **Xlist** to **L4** or whatever list holds the data and the frequency to **1.** Then press **GRAPH.** Press **TRACE** and press the right arrow to see the minimum, first quartile, median, third quartile, and maximum.

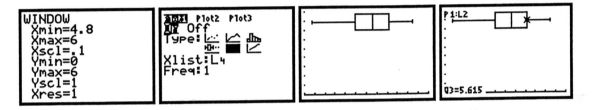

Statistics for Data in a Frequency Chart

Number of children per household. Compute the statistics and make a boxplot for the number of children living in a household.

Number of children	0	1	2	3	4	5	6
Number of households	60	42	86	59	22	4	2

Solution: First clear lists **L1** and **L2** in the **STAT Edit** screen, and then enter the numbers of children (measurements) into list **L1** and the numbers of households (frequencies) into list **L2.**

To compute the statistics, press **STAT,** scroll to **CALC,** and press **1.** Enter the command **1-Var Stats L1, L2,** which means that the measurements in list **L1** occur with frequency **L2.** If the data were stored in other lists, such as **L3** and **L4,** then we would have entered the command **1-Var Stats L3, L4.**

We see that there were 275 measurements and the average number of children per household was 1.858.

```
1-Var Stats
x̄=1.858181818
Σx=511
Σx²=1441
Sx=1.339284441
σx=1.336847161
↓n=275
```

```
1-Var Stats
↑n=275
minX=0
Q1=1
Med=2
Q3=3
maxX=6
```

For a boxplot, adjust the **WINDOW** to an appropriate range for **X** (the **Y** range can be ignored). Set **XList** and **Freq** in the **STAT PLOT** settings, and graph.

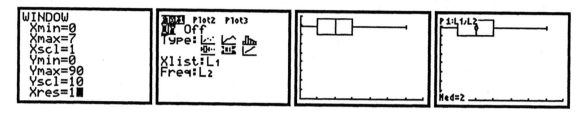

Computing Statistics for Two Data Sets of Common Size

If we have two data sets with an equal number of measurements (no more than 999), then we can enter the data into the **STAT EDIT** screen in order to compute the statistics simultaneously with the **2-Var Stats** command.

1.42 $\bar{x}$ and s are not enough. Evaluate $\bar{x}$ and s for the following data sets.

A	9.14	8.14	8.74	8.77	9.26	8.10	6.13	3.10	9.13	7.26	4.74
B	6.58	5.76	7.71	8.84	8.47	7.04	5.25	5.56	7.91	6.89	12.50

Solution: Enter data set **A** into list **L1** and data set **B** into list **L2**. To compute the statistics simultaneously, press **STAT**, press the right arrow to display the **CALC** screen, and press **2** for the command **2-Var Stats**. Then type L1, L2 and enter the resulting command **2-Var Stats L1, L2**. The screen initially shows the statistics from **L1**. Scroll down to see the statistics from **L2**.

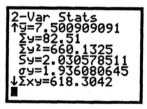

If the data were in other lists, say **L3** and **L4**, then we would enter the command **2-Var Stats L3, L4**.

If the data sets were of different sizes, then the statistics would have to be computed separately. In the previous exercise, we would enter the command **1-Var Stats L1** (from the **STAT CALC** screen) to find the statistics for set **A** and enter **1-Var Stats L2** to find the statistics for set **B**.

1.3 The Normal Distributions

One of the best statistical features on the TI-83 is the built-in normal distribution command in the **DISTR** menu.   For normally distributed measurements, with specified mean μ and standard deviation σ, we can find the area under the $N(\mu, \sigma)$ density curve between the values of a and b with the command **normalcdf(a, b, μ, σ)**.   In this section, we shall compute various normal probabilities, as well as "backward" normal calculations.

The Built-in Normal Distribution Command

1.58 How hard do locomotives pull?   The adhesion of one 4400-horsepower diesel locomotive model varies according to a normal distribution with mean $\mu = 0.37$ and standard deviation $\sigma = 0.04$.   What proportion of adhesions are (a) higher than 0.40?   (b) between 0.40 and 0.50?   (c) Improvements are made so that now $\mu = 0.41$ and $\sigma = 0.02$. Find the proportions in (a) and (b) after this improvement.

Solution: We can work part (b) directly with the built-in command. Press **2nd VARS (DISTR)**, then press 2. Then finish typing the command **normalcdf(0.40, 0.50, 0.37, 0.04)** and press **ENTER**. We see that around 22.6% of adhesions are between 0.40 and 0.50.

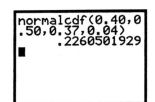

For part (a), we first note that 50% of adhesions are at least 0.37 (the value of the mean). This 50% breaks down into the amounts from 0.37 to 0.40 plus the amounts above .40. Thus, the proportion of adhesions above .40 is given by .50 *minus* the proportion from 0.37 to 0.40.

We enter the command **.50–normalcdf(0.37, 0.40, 0.37, 0.04)** and find that around 22.66% of the adhesions are above 0.40.

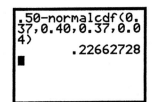

(c) To recompute (b) with the new mean and standard deviation, we enter the command **normalcdf(0.40, 0.50, 0.41, 0.02)**. We see that now over 69% of adhesions fall between 0.40 and 0.50.

Before we recompute the proportion in (a), we note that .40 is now *below* the mean of .41. Thus the amount above .40 will be the amount from .40 to .41 plus .50. So we must enter the command **normalcdf(0.40, 0.41, 0.41, 0.02) + .50**.

Later we shall see an alternate way to compute tail values without having to subtract from or add to .50.

The NORMDIST Program

To compute normal probabilities, we could also use a program that prompts the variables. Such a program is **NORMDIST**, and it is given on page 15. It must first be keyed or downloaded into the TI-83.

To execute the **NORMDIST** program for part (b) of Exercise 1.58, call up the program from the **PRGM** menu and press **ENTER**. Then enter **.37** for **MEAN**, enter **.04** for **ST DEV**, enter **.40** for **LOWER BOUND**, and enter **.50** for **UPPER BOUND**.

Then, if you wish to see a shaded graph of the normal curve, enter **1** for **GRAPH?**, otherwise enter **0**. If you enter **1**, then the graph of the partially shaded bell curve initially appears but then is replaced by the display of the computed probability. To see the graph again, press **GRAPH**. Then to see the probability value again, press **CLEAR**.

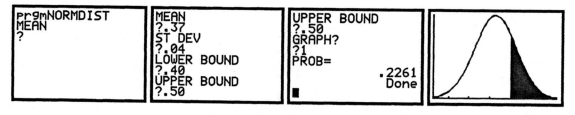

Again we see that around 22.6% of adhesions fall between .40 and .50.

```
PROGRAM:NORMDIST                :"normalpdf(X,M,S)"→Y₁
:Disp "MEAN"                     :M-3S→Xmin
:Input M                        :M+3S→Xmax
:Disp "ST DEV"                   :S→Xscl
:Input S                        :0→Ymin
:Disp "LOWER BOUND"              :Y₁(M)→Ymax
:Input J                        :.1→Yscl
:Disp "UPPER BOUND"              :If K>J
:Input K                        :Then
:If K>J                         :Shade(0,Y₁,J,K)
:Then                           :Else
:normalcdf(J,K,M,S)→B           :Shade(0,Y₁,⁻1E99,K)
:Else                           :End
:normalcdf(M,K,M,S)→B           :End
:End                            :If J=K
:round(B,4)→B                   :Then
:Disp "GRAPH?"                   :Disp "CUMULATIVE=",B+.5
:Input Z                        :Disp "RIGHT TAIL=",.5-B
:If Z=1                         :Else
:Then                           :Disp "PROB=",B
:PlotsOff                       :End
:FnOff
```

Computing Tail Values

A probability value such as $P(X \leq k)$ or $P(X < k)$ is called the *cumulative area* (or *left-tail value*). A probability value such as $P(X \geq k)$ or $P(X > k)$ is called the *right-tail value*. Both of these values can be computed simultaneously with the **NORMDIST** program, for any single value k, by entering the value of k for both the **LOWER BOUND** and the **UPPER BOUND**. In this case, the cumulative area will be shaded if entering **1** for **Graph?**.

An alternate way to compute a left-tail value $P(X \leq k)$ is with the command **normalcdf(⁻1 E 99, k, μ, σ)**. A right-tail value $P(X \geq k)$ can be found with the command **normalcdf(k, 1 E 99, μ, σ)**.

Example. Rework part (a) of Exercise 1.58 using the **NORMDIST** program and the alternate built-in command.

Solution: Call up the program from the **PRGM** menu, then enter **.37** for **MEAN**, enter **.04** for **ST DEV**, enter **.40** for **LOWER BOUND**, and enter **.40** again for **UPPER BOUND**. Enter **0** for **GRAPH?**. We see a **RIGHT TAIL** of .2266, so again 22.66% of adhesions are above .40.

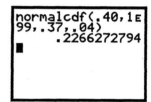

To use the built-in command for finding the amount above .40, we note that there is no theoretical upper bound. But we can use **1 E 99** (10^{99}) as an upper bound that is essentially infinite. So we enter the command **normalcdf(.40, 1 E 99, .37, .04)**.

1.65. Let $Z \sim N(0, 1)$ be the standard normal curve. Shade the areas and find the proportions for the regions (a) $Z \le -2.25$, (b) $Z \ge -2.25$, (c) $Z > 1.77$, (d) $-2.25 < Z < 1.77$.

Solution: For each part, we use the **NORMDIST** program, with a **MEAN** of 0 and a **ST DEV** of 1. We can do parts (a) and (b) together by entering **⁻2.25** for both bounds. To see the shaded region, enter **1** for **GRAPH?**.

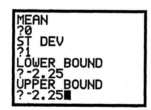 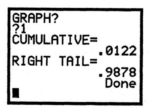 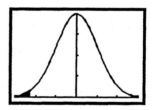

(a) We see that $P(Z \le -2.25) = .0122$ with a very small shaded area at the left tail of the graph. (b) We also see that $P(Z \ge -2.25) = .9878$, and its region is the large unshaded portion under the curve.

For part (c), we rerun the program and enter **1.77** for both bounds. For part (d), we enter bounds of **⁻2.25** and **1.77**. In both cases, we enter **1** for **GRAPH?**.

(c) We see that $P(Z > 1.77) = .0384$. In this case, the desired right tail is the unshaded portion of the graph.

(d) We see that $P(-2.25 < Z < 1.77) = .9494$.

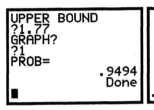

Alternately we can use the **ShadeNorm(** command from the **DISTR DRAW** menu. For part (c), enter **ShadeNorm(1.77, 1E99, 0, 1)**.

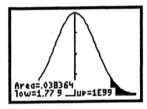

Alternately for part (a) enter **ShadeNorm(ˉ1E99, ˉ2.25, 0, 1)**. For part (b), enter **ShadeNorm(ˉ2.25, 1E99, 0, 1)**. For part (d), enter **ShadeNorm(ˉ2.25, 1.77, 0, 1)**.

1.88 Mexican Americans.   The ARSMA adaptation scores are normally distributed with a mean of 3.0 and a standard deviation of 0.8. What proportion of scores lies (a) below 1.7? (b) between 1.7 and 2.1?

Solution: We will use the built-in **normalcdf(** command from the **DISTR** menu.

We see that $P(X < 1.7) = .0521$, and $P(1.7 \leq X \leq 2.1) = .0782$.

```
normalcdf( -1E99,
1.7,3.0,0.8)
        .0520812696
normalcdf(1.7,2.
1,3.0,0.8)
        .0782132947
```

These values can be verified with the **NORMDIST** program.

"Backward" Normal Calculations

We often would like to know the value x for which $P(X \leq x)$ equals a desired proportion p. This "backward" value of x is formally called an *inverse cumulative normal value* and can be found with the **invNorm** command from the **DISTR** menu.

For example, for ARSMA scores with a mean of 3.0 and a standard deviation of 0.8, below what score do 90% of people fall?

Here the desired probability is .90, so we enter **invNorm (.90, 3.0, 0.8)** to obtain a value of 4.025241253.

1.89 The size of soldiers' heads. The head circumference among soldiers is normally distributed with a mean of 22.8 inches and a standard deviation of 1.1 inches. What sizes represent the smallest and largest 5%?

Solution: The smallest 5% is given by the command **invNorm(.05, 22.8, 1.1)**. The measurement for the largest 5% will be such that 95% fall *below* this value. Thus we use the command **invNorm(.95, 22.8, 1.1)**.

```
invNorm(.05,22.8
,1.1)
        20.99066101
invNorm(.95,22.8
,1.1)
        24.60933899
■
```

We find that 5% of soldiers measure no more than 20.99 inches, while 5% of soldiers measure 24.61 inches or more.

Assessing Normality

1.75 (d) Drive time. Assess the normality of Professor Moore's driving times to campus by finding the percentages that lie within one, two, and three standard deviations of average.

8.25	7.83	8.30	8.42	8.50	8.67	8.17	9.00	9.00	8.17	7.92
9.00	8.50	9.00	7.75	7.92	8.00	8.08	8.42	8.75	8.08	9.75
8.33	7.83	7.92	8.58	7.83	8.42	7.75	7.42	6.75	7.42	8.50
8.67	10.17	8.75	8.58	8.67	9.17	9.08	8.83	8.67		

Solution: First enter the data into a list, say **L1**. From the **STAT** menu, enter the command **SortA(L1)** to sort the data into increasing order.

Next compute the statistics by entering the command **1-Var Stats L1** from the **STAT CALC** screen. As before, we must now access the statistics from the **VARS Statistics** menu to compute the ranges for one, two, and three standard deviations from average. Compute these ranges as shown.

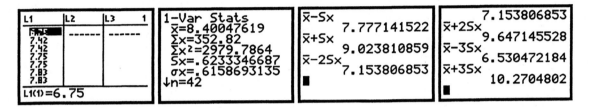

Now press **STAT** and press **1** to return to the list. Scrolling down the list and counting, we see that 33 out of 42, or 78.57%, of the measurements lie within one standard deviation of average from 7.777 to 9.024. We see that 39 out of 42, or 92.86%, of the measurements lie between 7.1538 and 9.647. Finally we see that 100% of the measurements are within three standard deviations of average. These results are slightly different from the 68-95-99.7 rule.

CHAPTER
2

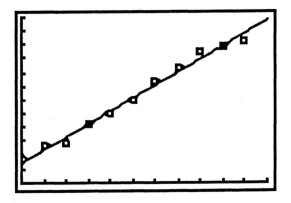

Examining Relationships

In this chapter we graph the relationship between two quantitative variables using a scatterplot. We then study the correlation and find the least-squares regression line. We also study the resulting residuals of this linear fit. Last, we provide a program that converts a two-way table of raw data into three types of proportion tables.

2.1 Scatterplots

We first plot two quantitative variables along the x and y axes to see if we can observe a relationship. In particular, we look for a linear relationship.

2.6 Does fast driving waste fuel? Make a scatterplot of the fuel used (liters/100 km) versus speed (km/h).

Speed	Fuel used	Speed	Fuel used	Speed	Fuel used
10	21.00	60	5.90	110	9.03
20	13.00	70	6.30	120	9.87
30	10.00	80	6.95	130	10.79
40	8.00	90	7.57	140	11.77
50	7.00	100	8.27	150	12.83

Solution: We first enter the data into the **STAT Edit** screen. Clear (or store into other lists) any data from **L1** and **L2**. Enter the speed into **L1** and the fuel used into **L2**. The speed will be plotted on the x axis and the fuel used will be plotted on the y axis. Adjust the **WINDOW** as shown so that the ranges include all measurements. Adjust the **STAT PLOT** settings by highlighting and entering the first **Type** and setting the appropriate lists. Press **GRAPH** to see the scatterplot, and then press **TRACE** if so desired.

We can observe that very low speeds and very high speeds cause the greatest fuel consumption.

2.7 Do heavier people burn more energy? Make a scatterplot of mass versus metabolic rate for the females. Make another scatterplot with a different symbol for the males, and then combine the two plots.

Sex	Mass	Rate	Sex	Mass	Rate
M	62.0	1792	F	40.3	1189
M	62.9	1666	F	33.1	913
F	36.1	995	M	51.9	1460
F	54.6	1425	F	42.4	1124
F	48.5	1396	F	34.5	1052
F	42.0	1418	F	51.1	1347
M	47.4	1362	F	41.2	1204
F	50.6	1502	M	51.9	1867
F	42.0	1256	M	46.9	1439
M	48.7	1614			

Solution: We first enter the mass and rate of just the females into lists **L1** and **L2** respectively. Then we enter the mass and rate of the males into lists **L3** and **L4**. However, we adjust the **WINDOW** so that the **X** range includes all the masses and the **Y** range includes all the rates. We adjust the **STAT PLOT** settings in **Plot1** as in the previous exercise to obtain the scatterplot of **L1** versus **L2**.

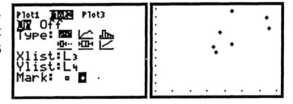

Next turn off **Plot1**, turn on **Plot2**, and adjust its settings with a different **Mark** for the males, and plot **L3** versus **L4**.

To see the females and males together, turn on both **Plot1** and **Plot2** and regraph.

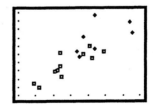

2.16 Categorical explanatory variable. Make a plot of the counts of insects trapped against board color. Compute the means for each color, add the means to the plot, and connect the means with a line segment.

Board color	Insects trapped					
Lemon yellow	45	59	48	46	38	47
White	21	12	14	17	13	17
Green	37	32	15	25	39	41
Blue	16	11	20	21	14	7

Solution: We will plot the colors on the x axis consecutively as the values 1, 2, 3, and 4. Since there are six measurements for each color, we enter each of the values 1 through 4 a total of six times into list **L1** as shown below. We list the corresponding measurements in **L2**.

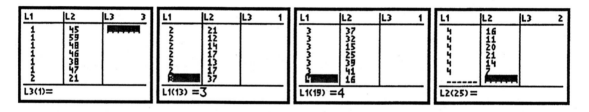

Next we adjust the **WINDOW** so that the **X** range includes the values 1 through 4 and the **Y** range includes all the measurements. Adjust the **STAT PLOT** settings and graph.

Next, we put the values 1 through 4 into list **L3** and then compute the mean number of insects trapped for each color; we put the means into list **L4**. (Since there are only six measurements, it is easy to compute the means "by hand" by summing and dividing by 6.) Turn on **Plot2**, set it to the second **Type**, and set the lists to **L3** and **L4**. Graph **Plot1** and **Plot2** together to see the measurements and the means.

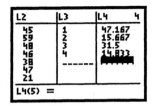
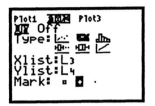
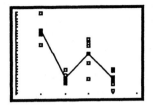

2.2 Correlation

In this section, we use the TI-83 to compute the correlation coefficient r and the squared correlation coefficient r^2 between paired data of quantitative variables.

In order to compute the correlation, we must first make sure that the calculator's diagnostics are turned on. To turn the setting on, press **2nd 0** (**CATALOG**) and scroll down to the **DiagnosticOn** command. Press **ENTER** to bring the command to the Home screen, then press **ENTER** again.

2.17 Classifying fossils.   Make a scatterplot, then compute the correlation coefficient r between lengths (in centimeters) of these preserved bones of five specimens of *archaeopteryx*.

Femur	38	56	59	64	74
Humerus	41	63	70	72	84

Solution: We first enter the data into lists **L1** and **L2**. Clear any data from these lists and enter the femur lengths into **L1** and the humerus lengths into **L2**. Adjust the **WINDOW** and **STAT PLOT**, then graph.

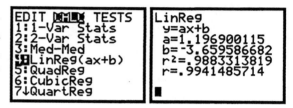

To find the correlation, press **STAT**, scroll right to **CALC**, press **4**, then press **ENTER**. We see that $r \approx .994$, which depicts the strong linear relationship.

Note: Regression defaults are for lists **L1** and **L2**. If the data were in other lists, say lists **L3** and **L4**, then we would first press **4** to obtain the command **LinReg(ax+b)** on the Home screen, and then we would enter the command **LinReg(ax+b) L3, L4**.

2.23 How many calories?   Find the correlation between the average guessed calories and the correct calories of these foods. Then remove the spaghetti and snack cake and find the correlation for the remaining foods.

Food	Guessed calories	Correct calories	Food	Guessed calories	Correct calories
8 oz. milk	196	159	candy bar	364	260
spaghetti	394	163	cracker	74	12
macaroni	350	269	apple	107	80
wheat bread	117	61	potato	160	88
white bread	136	76	snack cake	419	160

Solution: We will enter this data into lists **L3** and **L4**. After doing so, press **STAT**, scroll right to **CALC**, press **4** to obtain the command **LinReg(ax+b)** on the Home screen, and then enter the command **LinReg(ax+b) L3, L4**.

So $r \approx .8245$. To delete the two foods, press **STAT**, press **1**, scroll down list **L3** and highlight the **394**, then press **DEL**. Scroll down to highlight **419** and press **DEL**. Then delete the two corresponding measurements from **L4**. Press **2nd QUIT** to return to the Home screen. Now press **2nd ENTER** to retrieve the previous command **LinReg(ax+b) L3, L4** and press **ENTER**.

By deleting the two foods, we find that the correlation has now increased to $r \approx .98374$.

```
LinReg
 y=ax+b
 a=.8435633126
 b=-32.96490278
 r²=.9677467252
 r=.9837411881
```

2.3 Least-Squares Regression

In this section we compute the least-squares line of two quantitative variables and graph it through the scatterplot of the variables. We then use the line to predict the y value that should occur for a given x measurement.

2.33 The professor swims. Here are Professor Moore's times (in minutes) to swim 2000 yards and his pulse rate afterwards (in beats per minute). (a) Make a scatterplot of the data, then calculate and graph the least-squares regression line. (b) Predict the professor's pulse rate for a time of 34.30 minutes. (c) Predict the time for a pulse rate of 152.

Time	34.12	35.72	34.72	34.05	34.13	35.72	36.17	35.57
Pulse	152	124	140	152	146	128	136	144
Time	35.37	35.57	35.43	36.05	34.85	34.70	34.75	33.93
Pulse	148	144	136	124	148	144	140	156
Time	34.60	34.00	34.35	35.62	35.68	35.28	35.97	
Pulse	136	148	148	132	124	132	139	

Solution: We make a scatterplot in the usual way. Here we will use lists **L5** (times) and **L6** (pulse rates). Adjust the **WINDOW** and **STAT PLOT** settings, and then graph.

L4	L5	L6	4
159	34.12	152	
269	35.72	124	
61	34.72	140	
76	34.05	152	
260	34.13	146	
12	35.72	128	
80	36.17	136	
L4 =(159,269,61,...			

```
WINDOW
Xmin=-5
Xmax=40
Xscl=5
Ymin=50
Ymax=900
Yscl=50
Xres=1
```

```
Plot1 Plot2 Plot3
 On Off
Type: ▦ ⊠ ⊪
        ⊪ ⊞ ⊠
Xlist:L5
Ylist:L6
Mark: □  +  ·
```

Regression Line of y on x

To obtain the linear regression line, press **STAT**, scroll right to **CALC**, press **4**, to obtain the command **LinReg(ax+b)** on the Home screen, and then enter the command **LinReg(ax+b) L5, L6**.

We see that the equation of the line is $y = -9.6949x + 479.934$, where x is the time.

The values of r and r^2 are also given.

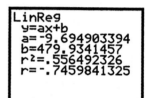

```
LinReg
 y=ax+b
 a=-9.694903394
 b=479.9341457
 r²=.556492326
 r=-.7459841325
```

Note: To obtain the equation in the form $y = a + bx$, use the **LinReg(a+bx)** command (item **8** in **STAT CALC**) and enter **LinReg(a+bx) L5, L6**. The result will be $y = 479.934 - 9.6949x$.

To graph the line, we must enter it into the **Y=** screen. We can type it directly, or we can access this regression equation from the **VARS Statistics** menu. Press **Y=** and clear any function that might be in **Y1**. (Make sure the calculator is on **Func** mode.) Now press **VARS**, then press **5** for **Statistics**. Scroll right to **EQ**, and press **1**. The regression equation is entered into **Y1**. Press **GRAPH**.

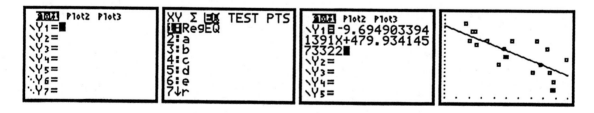

Prediction

To evaluate the line at a specified x, we now access **Y1** from the **Y-VARS** screen. Press **VARS**, scroll right to **Y-VARS**, press **1**, then press **1** again. The function **Y1** is entered into the Home screen. To evaluate the pulse rate for a time of $x = 34.30$, enter the command **Y1(34.30)**.

We see the predicted pulse rate of just over 147.

We can also verify that the point $(\bar{x}, \bar{y})$ is on the line. First we compute the statistics. We can do so simultaneously with the **2-Var Stats** command from the **STAT CALC** menu since the two data sets have the same number of points. Enter the command **2-Var Stats L5, L6**. Then enter **Y1($\bar{x}$)** by recalling $\bar{x}$ from the **VARS Statistics** menu.

We see that
Y1($\bar{x}$) = $\bar{y}$.

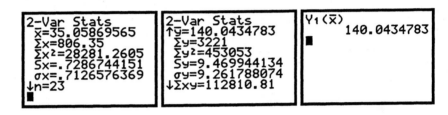

Regression Line of x on y

For part (c) of Exercise 2.33 we are asked to predict an x value (time) given a specific y value (pulse rate). To do so, we can find the regression line of x on y and then evaluate it at the given pulse rate.

To compute the least-squares regression line for pulse rates on times, enter the command **LinReg(ax+b) L6, L5**. Now enter this regression equation into **Y2** as we did for **Y1**. Then return to the Home screen and enter **Y2(152)** to find the predicted time for a pulse rate of 152.

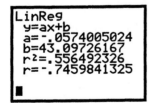

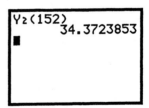

We can predict a time of 34.37 minutes, fairly close to the first sample time of 34.12 minutes, which actually did result in a pulse rate of 152.

2.4 Cautions about Correlation and Regression

We now provide an exercise to demonstrate how to plot the residuals of a least-squares line.

2.52 Predicting enrollments. The data below give the numbers of entering freshmen x and the numbers of students y taking freshman-level math at a large university. Make a scatterplot, compute and graph the least-squares regression line, and then plot the residuals against the year.

Year	1991	1992	1993	1994	1995	1996	1997	1998
x	4595	4827	4427	4258	3995	4330	4265	4351
y	7364	7547	7099	6894	6572	7156	7232	7450

Solution: We first enter the years into list **L1** and the corresponding measurements into lists **L2** and **L3**. We then adjust the **WINDOW** and **STAT PLOT** settings to graph L3 versus L2.

We next compute the regression line with the command **LinReg(ax+b) L2, L3,** enter it into **Y1,** and regraph.

We see that $r^2 \approx .6943$; thus, around 69.43% of the variation in class enrollment is explained by the linear relationship with freshmen enrollment.

Residual Plot

After computing a regression, the residuals are stored in the **LIST** screen. Press **2nd STAT (LIST)**, and press **1** to obtain the command **LRESID** on the Home screen. Enter the command **LRESID →L4** to store the residuals in **L4**.

Enter the **STAT Edit** screen and scroll down list **L4** to decide upon an appropriate range for the residuals. Set the **X** range in the **WINDOW** to include all the years and set the **Y** range to include all the residuals. Adjust the **STAT PLOT** to graph **L4** versus **L1**. Deselect the regression line **Y1** in the **Y=** screen, then press **GRAPH** to plot the residuals against the year.

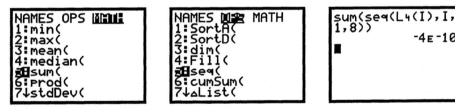

Sum of Residuals

We can also verify that the residuals sum to 0 (up to round-off). Press **2nd STAT**, scroll right to **MATH** and press **5** to bring the command **sum(** to the Home screen. Next press **2nd STAT**, scroll right to **OPS**, and press **5** for the command **seq(** to make the command on the Home screen become **sum(seq(**. Now we wish to sum the eight terms in list **L4**. Complete typing the command **sum(seq(L4(I), I, 1, 8))** and press **ENTER**.

We see that the sum of the residuals is 0 up to round-off error.

2.5 Relations in Categorical Data

We now provide a supplemental program that converts a two-way table of raw data into three different proportion tables. The **TWOWAY** program given on page 33 will give the proportion of each category among the whole population, the conditional distributions of each column category, and the conditional distributions of each row category.

2.74 Majors for men and women in business. In a study of the career plans of students in the College of Business Administration at the University of Illinois, one question asked which major the student had chosen. The responses follow.

	Female	Male
Accounting	68	56
Administration	91	40
Economics	5	6
Finance	61	59

(a) Compute the proportions of responses in each category.
(b) Find the conditional distributions for women and for men.
(c) Find the conditional distributions for each of the four majors.

Solution: We can compute all three types of proportions with the **TWOWAY** program. The data set creates a 4 × 2 matrix. There are 4 rows (types of major) and 2 columns (sex). Before running the program, we must enter these data into matrix **[A]** in the **MATRX EDIT** screen.

Press **MATRX**, scroll right to **EDIT**, and press **1**. Enter the dimensions for matrix **[A]** as 4 × 2. Then enter the data into the matrix.

```
NAMES MATH EDIT
1:[A]
2:[B]
3:[C]
4:[D]
5:[E]
6:[F]
7↓[G]
```

```
MATRIX[A] 4 ×2
[ 0        ]
[ 0        ]
[ 0        ]
[ 0        ]
```

```
MATRIX[A] 4 ×2
[ 68    56  ]
[ 91    40  ]
[ 5     6   ]
[ 61    59  ]
```

```
PROGRAM:TWOWAY
:Disp "NO. OF ROWS"
:Input R
:Disp "NO. OF COLUMNS"
:Input C
:{R+1,C+1}→dim([B])
:0→N
:For(I,1,R)
:N+sum(seq([A](I,J),J,1,C))→N
:End
:For(I,1,R)
:For(J,1,C)
:round([A](I,J)/N,4)→[B](I,J)
:End
:End
:For(I,1,R)
:round(sum(seq([A](I,J),J,1,C))/N,4)→[B](I,C+1)
:End
:For(J,1,C)
:round(sum(seq([A](I,J),I,1,R))/N,4)→[B](R+1,J)
:End
:1→[B](R+1,C+1)
:{R+1,C}→dim([C])
:For(J,1,C)
:sum(seq([A](I,J),I,1,R))→A
:For(I,1,R)
:round([A](I,J)/A,4)→[C](I,J)
:End
:1→[C](R+1,J)
:End
:{R,C+1}→dim([D])
:For(I,1,R)
:sum(seq([A](I,J),J,1,C))→A
:For(J,1,C)
:round([A](I,J)/A,4)→[D](I,J)
:End
:1→[D](I,C+1)
:End
```

Now press **PRGM**, then scroll down and call up the **TWOWAY** program. Enter the dimensions into program.

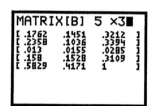

```
PrgmTWOWAY
NO. OF ROWS
?4
NO. OF COLUMNS
?2
```

The program computes the desired proportion tables and stores them into matrices **[B]**, **[C]**, and **[D]**. Press **MATRX**, then press **2** to call up matrix **[B]**. Scroll right to see the remainder of the matrix on the screen.

```
[B]
[[.1762 .1451 ...
 [.2358 .1036 ...
 [.013  .0155 ...
 [.158  .1528 ...
 [.5829 .4171 1...
```

```
[B]
...2 .1451 .3212]
...8 .1036 .3394]
...  .0155 .0285]
...  .1528 .3109]
...9 .4171 1    ]]
```

(You can also view matrix **[B]** from within the **MATRX EDIT** screen.)

```
MATRIX[B] 5 ×3█
[ .1762   .1451   .3212 ]
[ .2358   .1036   .3394 ]
[ .013    .0155   .0285 ]
[ .158    .1528   .3109 ]
[ .5829   .4171   1     ]
```

Matrix **[B]** gives the overall proportions from the entire sample. Notice that matrix **[B]** is now 5 × 3. The totals are in the last row and column. Of the respondents, 58.29% were female and 41.71% were male (from the last row). From the last column, we see that 32.12% were accounting majors, while only 2.85% were economics majors. We also see that 17.62% of all respondents were female accounting majors while 14.51% were male accounting majors.

Next observe matrices **[C]** and **[D]** by calling them up from the **MATRX** menu.

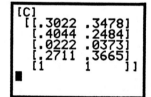

```
[C]
[[.3022 .3478]
 [.4044 .2484]
 [.0222 .0373]
 [.2711 .3665]
 [1     1    ]]
█
```

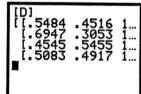

```
[D]
[[.5484 .4516 1...
 [.6947 .3053 1...
 [.4545 .5455 1...
 [.5083 .4917 1...
█
```

Matrix **[C]** gives the conditional distribution of majors among females and among males. Among females, 30.22% were accounting majors, 40.44% were administration majors, 2.22% were economics majors, and 27.11% were finance majors. Among males, 34.78% were accounting majors, 24.84% were administration majors, 3.73% were economics majors, and 36.65% were finance majors.

Now consider the values in matrix **[D]**. Among accounting majors, 54.84% were female and 45.16% were male. Among administration majors, 69.47% were female and 30.53% were male. Among economics majors, 45.45% were female and 54.55% were male. Among finance majors, 50.83% were female and 49.17% were male.

2.86 Age and marital status of women.   From the following data set on age and marital status of women in 1995, analyze the three possible proportion tables.

Marital status

Age	Never married	Married	Widowed	Divorced	Total
18–24	9289	3046	19	260	12,613
25–39	6948	21,437	206	3408	32,000
40–64	2307	26,679	2219	5508	36,713
≥ 65	768	7767	8636	1091	18,264
Total	19,312	58,931	11,080	10,266	99,558

Solution: The totals are not necessary when entering the data. Press **MATRX**, scroll right to **EDIT**, and press 1. Enter the dimensions for matrix **[A]** as 4 × 4. Then enter the data into the matrix.

```
MATRIX[A]  4 ×4
[ 9289    3046    19    -
[ 6948    21437   206   -
[ 2307    26679   2219  -
[ 768     7767    8636  -
```

Execute the **TWOWAY** program. Then call up matrix **[B]** on the Home screen (or view it in the **MATRX EDIT** screen). Scroll right to see the remainder of the matrix.

```
[B]
[[.0933 .0306 2…
[.0698 .2153 .…
[.0232 .2679 .…
[.0077 .078  .…
[.1939 .5917 .…
```

```
[B]
… 2E-4  .0026 .…
… .0021 .0342 .…
… .0223 .0553 .…
… .0867 .011  .…
… .1113 .1031 1…
```

```
[B]
…   .0026 .1267]
…1 .0342 .3213]
…3 .0553 .3686]
…7 .011  .1834]
…3 .1031 1     ]]
```

Matrix **[B]** gives the proportions among the entire sample that was surveyed. By looking at the last row, we see that 19.39% of women were never married, 59.17% were married, 11.13% were widowed, and 10.31% were divorced.

From the last column, we see that 12.67% of women were age 18–24, while 32.13% were age 25–39. Looking at column 1, entry 4, we see that 7.7% were 65 or older and never married. From row 2, entry 2, we see that 21.53% of women were age 25–39 and married.

Now let's look at matrix **[C]** and analyze the second column.

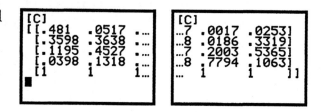

Among married women, 5.17% were age 18–24, 36.38% were age 25–39, 45.27% were age 40–64, and 13.18% were 65 or older.

Finally, let's look at matrix **[D]** and analyze the third row.

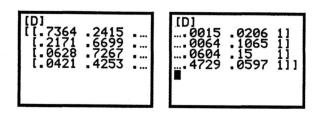

Among women age 40–64, 6.28% were never married, 72.67% were married, 6.04% were widowed, and 15% were divorced.

CHAPTER
3

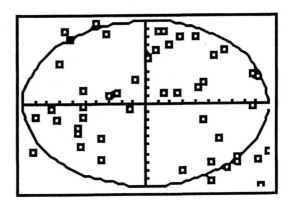

Producing Data

In this chapter we provide supplementary TI-83 programs that simulate the gathering of random samples. The programs can be used to study the properties of sampling statistics.

3.1 Designing Samples

The first program, **SAMPLEP**, generates count data, or *Bernoulli trials*, for a specified proportion p. The second program, **SAMPLEN**, generates random data from a specified normal distribution. After keying in or downloading the programs, we execute them by specifying the parameters and the desired sample size. The programs enter the random data into list **L1** and then display a comparison between the actual parameters and the sample statistics.

Restriction: We can specify a sample size of at most 999.

```
PROGRAM:SAMPLEP              PROGRAM:SAMPLEN
:Disp "REAL P"              :Disp "MEAN"
:Input P                    :Input M
:Disp "SAMPLE SIZE"         :Disp "ST DEVIATION"
:Input N                    :Input S
:ClrList L₁                 :Disp "SAMPLE SIZE"
:For(I,1,N)                 :Input N
:If rand≤P                  :randNorm(M,S,N)→L₁
:Then                       :1-Var Stats L₁
:1→L₁(I)                    :Disp "REAL MEAN,DEV"
:Else                       :Disp M,S
:0→L₁(I)                    :Disp "SAMPLE MEAN,DEV"
:End                        :Disp x̄,Sx
:End
:1-Var Stats L₁
:Disp "REAL P"
:Disp P
:Disp "SAMPLE PROP."
:Disp x̄
```

Here are the results of one running of the **SAMPLEP** program with a proportion of $p = .72$ and a sample size of $n = 60$.

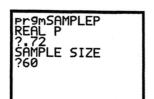
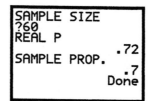
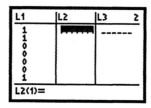

The program chose 60 consecutive random numbers between 0 and 1. If a number was less than or equal to .72, then a **1** was entered into list **L1**. Otherwise a **0** was entered into the list. The result was that 70% of the random numbers were less than or equal to .72, which compares favorably to the true proportion of .72.

Here are the results of one running of the **SAMPLEN** program with a mean of 100, a standard deviation of 15, and a sample size of 80.

```
PrgmSAMPLEN
MEAN
?100
ST DEVIATION
?15
SAMPLE SIZE
?80■
```

```
REAL MEAN,DEV
              100
               15
SAMPLE MEAN,DEV
     97.82013403
     14.41419777
             Done
```

```
L1      L2      L3      2
83.749 ■■■■■■   ------
133.53
112.4
112.14
93.359
91.92
62.9
L2(1)=
```

The program chose 80 random measurements from a $N(100, 15)$ distribution and entered them into **L1**. The sample mean $\bar{x}$ and sample deviation s of these measurements were respectively 97.82 and 14.414.

Goodness of Fit

At the end of Section 1.3, we saw the 68-95-99.7 rule, which is an informal way to assess normality. We shall now look a little more closely at how well randomly generated normal data fit the theoretical normal curve. We will sort such data into blocks and make a histogram, then compute the frequencies in these blocks and compare them with the actual proportions that should occur under the normal curve.

Exercise. Generate 50 random sample points from the $N(20, 4)$ distribution. Divide the range $[\mu - 3\sigma, \mu + 3\sigma]$ into eight equidistant blocks. Compare the frequencies of the data occurring in these blocks with the theoretical proportions that should occur.

Solution: We first execute the **SAMPLEN** program with a mean of 20, a standard deviation of 4, and a sample size of 50. The generated sample points are stored in list **L1**. Observe the points and then clear lists **L2** and **L3**.

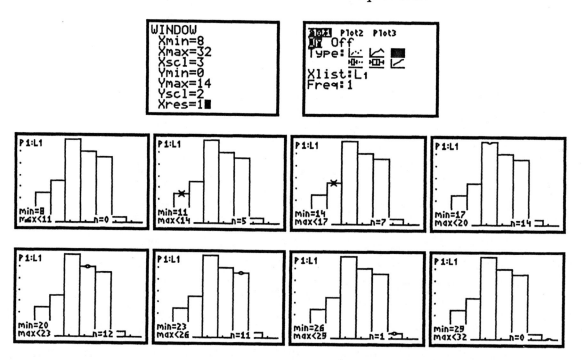

For the $N(20, 4)$ distribution, the range $[\mu - 3\sigma, \mu + 3\sigma]$ becomes $[8, 32]$, which has length 24. If we want eight equidistant blocks, then each block must have length 3. So adjust the **WINDOW** with **X** ranging from 8 to 32 on a scale of 3. The **Y** values must range from 0 to a height that allows the viewing of the entire histogram. Next, adjust the **STAT PLOT** settings to a histogram for **L1** occurring with frequency 1. Press **GRAPH**, and then press **TRACE**. Scroll across each block and note the frequencies.

For this running of the program, we see that the first block $[8, 11)$ has 0 data points; the next block $[11, 14)$ has 5 data points; the block $[14, 17)$ has 7 points. The list of frequencies is $\{0, 5, 7, 14, 12, 11, 1, 0\}$.

We now enter the *relative frequencies* into list **L2**. Since there were 50 data points, enter the values 0/50, 5/50, 7/50, 14/50, 12/50, 11/50, 1/50, 0/50 into list **L2**. You can type them as fractions, and the calculator will convert them into decimals when you press **ENTER**.

We now enter the true normal proportions that should occur in each block into list **L3**. Press **2nd QUIT** to return to the Home screen. Press **2nd VARS (DISTR)**, then press **2** to call up the command `normalcdf(`. To find the proportion for the first block and to store it as the first entry in L3, enter the command `normalcdf(8,11,20,4)`→**L3(1)**. Press **2nd ENTER** to recall the command, then edit it to **normalcdf(11,14,20,4)**→**L3(2)** and press **ENTER**.

Continue pressing **2nd ENTER** and editing the command for successive blocks.

```
normalcdf(8,11,2
0,4)→L3(1)
           .0108744662
normalcdf(11,14,
20,4)→L3(2)
           .0545827953
```

```
           .0545827953
normalcdf(14,17,
20,4)→L3(3)
           .1598200507
normalcdf(17,20,
20,4)→L3(4)
           .2733727211
■
```

```
           .2733727211
normalcdf(20,23,
20,4)→L3(5)
             .27337272
normalcdf(23,26,
20,4)→L3(6)
           .1598200507
```

Alternately, we can use the symmetry of the normal curve after the proportions for the first four blocks have been computed and stored. The fifth block has the same proportion as the fourth block; the sixth block has the same as the third; the seventh has the same as the second; and the eighth has the same as the first. After the proportions for the first four blocks are computed, we can use the commands below to finish the entries in list L3.

```
L3(4)→L3(5)
           .2733727211
L3(3)→L3(6)
           .1598200507
L3(2)→L3(7)
           .0545827953
L3(1)→L3(8)
```

Now press **STAT** and press **1** to view the lists. Scroll down lists **L2** and **L3** and compare the frequencies that did occur with the theoretical proportions that should occur. In this example, we see a slight anomaly in the sixth block.

L1	L2	L3	3
18.041	0	.0108
23.055	.1	.05458
24.927	.14	.15982
20.595	.28	.27337
25.6	.24	.27337
23.746	.22	.15982
22.092	.02	.05458

L3(6) =.15982

There are more formal tests, such as a chi-square test, to assess the "goodness of fit" of sample data to a theoretical distribution. Such tests allow one to decide whether anomalies are due to chance or due to the fact that the sample data really do not fit the expected distribution.

3.2 Designing Experiments

We now expand the previous programs to simulate the collection of several random samples. The program **SAMPLEP2** generates a specified number of random samples of count data, each of the same specified sample size and for the same proportion p. It computes and stores the sample proportion for each sample. Then it computes the average of the sample proportions to compare with the real proportion p.

The program **SAMPLEN2** generates a specified number of random samples, each of the same specified sample size from a specified normal distribution. It computes and stores the sample mean for each sample. Then it computes the average and standard deviation of the sample means to compare with the theoretical mean and standard deviation of the sampling statistics.

Restrictions: We can specify a sample size of at most 999 and at most 999 samples. (Memory and time will probably prevent using extremely large sample sizes or a large number of samples.)

Program:SAMPLEP2	
:Disp "REAL P"	:Else
:Input P	:0→L$_1$(I)
:Disp "SAMPLE SIZE"	:End
:Input N	:End
:Disp "NUM. OF SAMPLES"	:1-Var Stats L$_1$
:Input M	:x̄→L$_2$(J)
:ClrList L$_2$	:End
:For(J,1,M)	:1-Var Stats L$_2$
:ClrList L$_1$	:Disp "REAL P"
:For(I,1,N)	:Disp P
:If rand≤P	:Disp "AVE OF SAMPLE P"
:Then	:Disp x̄
:1→L$_1$(I)	

Here are the results of one running of the **SAMPLEP2** program with a proportion of $p = .36$, a sample size of 30, and 20 samples.

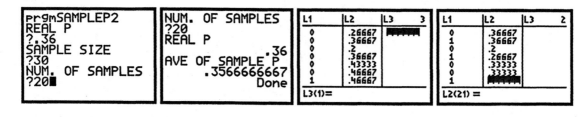

The program chose 30 consecutive random numbers between 0 and 1. If a number was less than or equal to .36, then a **1** was entered into list **L1**. Otherwise a **0** was entered into the list. The sample proportion was then computed and stored as the first entry in list **L2**. The procedure was repeated 19 more times with successive sample proportions stored in **L2**. The last random sample remains in list **L1** after the program ends.

Observing the list of sample proportions in **L2**, we see that we may never have a sample proportion that equals the real proportion of .36. However, the *average* of all 20 sample proportions was .3567, which is very close to .36.

```
Program:SAMPLEN2                    :randNorm(M,S,N)→L₁
:Disp "MEAN"                        :1-Var Stats L₁
:Input M                            :x̄→L₂(J)
:Disp "ST DEVIATION"                :End
:Input S                            :1-Var Stats L₂
:Disp "SAMPLE SIZE"                 :Disp "MEAN,σx/√(N)"
:Input N                            :Disp M,S/√(N)
:Disp "NUM. OF SAMPLES"             :Disp "SAMPLING STATS"
:Input K                            :Disp x̄,Sx
:ClrList L₂
:For(J,1,K)
```

Here are the results of one running of the **SAMPLEN2** program with a mean of 68, a standard deviation of 3, a sample size of 30, and 20 samples.

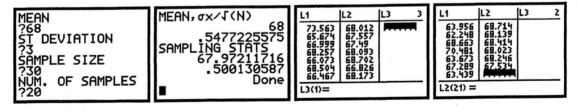

The program chose 30 random measurements from the $N(68, 3)$ distribution and entered them into **L1**. The sample mean was stored as the first entry in list **L2**. The procedure was repeated 19 more times with successive sample means stored in **L2**. The last random sample remains in list **L1** after the program ends.

Observing the list of sample means in **L2**, we see that we may never have a sample mean that equals the real mean of 68. But the average of all 20 sample means was 67.972.

The standard deviation of all possible sample means is theoretically given by $\sigma/\sqrt{n}$, where σ is the true population standard deviation and n is the sample size. Here, $\sigma/\sqrt{n} = 3/\sqrt{30} \approx .5477$. In this running of the program, the standard deviation of the 20 sample means was .50.

More Goodness of Fit

When choosing random samples of size n from a $N(\mu, \sigma)$ distribution, the set of all possible sample means will also be normally distributed. However, the sample means follow a $N(\mu, \sigma/\sqrt{n})$ distribution. There is less variance in the sample means than in the measurements. In other words, the sample means are consistently closer to the mean μ.

We will once again generate a collection of sample means using the **SAMPLEN2** program. As in Section 3.1, we will sort the sample means into blocks and make a histogram, then compute the frequencies in these blocks. But now we will compare the frequencies with the proportions that should occur under the $N(\mu, \sigma/\sqrt{n})$ curve.

Exercise. Generate 50 random sample means from random samples of size 25 from the $N(100, 15)$ distribution. Divide the range $[\mu - 3\sigma/\sqrt{n}, \ \mu + 3\sigma/\sqrt{n}]$ into nine equidistant blocks. Compare the frequencies of the sample means occurring in these blocks with the theoretical proportions that should occur.

Solution: We first execute the **SAMPLEN2** program with a mean of 100, a standard deviation of 15, a sample size of 25, and 50 samples. (Give it a couple of minutes to run!) The 50 sample means are stored in list **L2**. The results follow. View the results and also clear lists **L3** and **L4**.

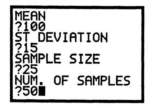

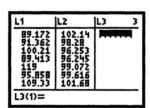

For the $N(100, 15)$ distribution and samples of size of $n = 25$, then $3\sigma/\sqrt{n} = 9$. Thus the range $[\mu - 3\sigma/\sqrt{n}, \mu + 3\sigma/\sqrt{n}]$ becomes [91, 109], which has length 18. If we want nine equidistant blocks, then each block must have length 2. So adjust the **WINDOW** with **X** ranging from 91 to 109 on a scale of 2. The **Y** values must range from 0 to a height that allows a view of the entire histogram. Next, adjust the **STAT PLOT** settings to a histogram for **L2** occurring with frequency 1. Press **GRAPH** and then press **TRACE**. Scroll across each block and note the frequencies.

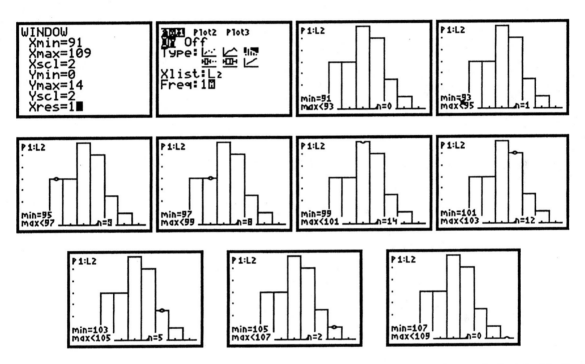

For this running of the program, the list of frequencies was {0, 1, 8, 8, 14, 12, 5, 2, 0}.

Now enter the relative frequencies into list **L3**. Since there were 50 data points, enter the values 0/50, 1/50, 8/50, 8/50, 14/50, 12/50, 5/50, 2/50, 0/50 into list **L3**. You can type them as fractions, and the calculator will convert them into decimals when you press **ENTER**.

L2	L3	L4	3
102.14	0	------	
98.28	.02		
96.253	.16		
96.245	.16		
99.072	.28		
99.616	.24		
101.68			

L3(7) =.1

Last, we wish to enter the true normal proportions from the $N(\mu, \sigma/\sqrt{n}) = N(100, 3)$ curve that should occur into list **L4**. Press **2nd QUIT** to return to the Home screen. Press **2nd VARS (DISTR)**, then press **2** to call up the command **normalcdf(** .

To find the proportion for the first block and to store it as the first entry in **L4**, enter the command **normalcdf(91,93,100,3)→L4(1)**. Press **2nd ENTER** to recall the command, then edit it to **normalcdf(93,95,100,3)→L4(2)** and press **ENTER**. Continue pressing **2nd ENTER** and editing the command for the first five blocks.

By symmetry of the normal curve, the last four blocks are equal to the first four blocks. We can either continue using the **normalcdf(** command for the last four blocks or use the alternate commands shown on the right.

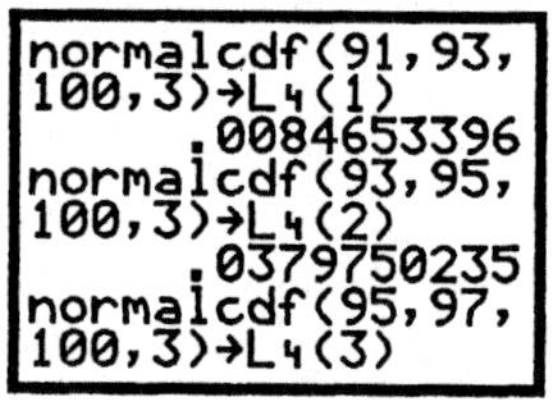

```
normalcdf(91,93,
100,3)→L4(1)
       .0084653396
normalcdf(93,95,
100,3)→L4(2)
       .0379750235
normalcdf(95,97,
100,3)→L4(3)
```

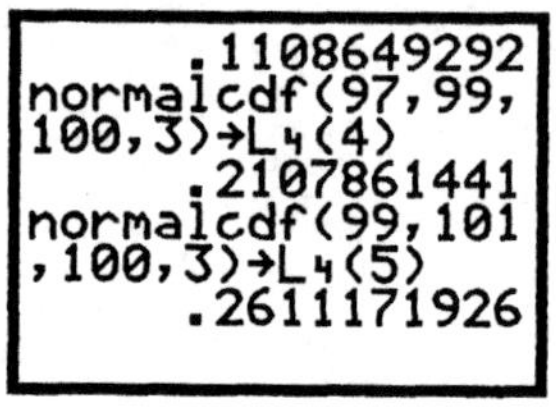

```
       .1108649292
normalcdf(97,99,
100,3)→L4(4)
       .2107861441
normalcdf(99,101
,100,3)→L4(5)
       .2611171926
```

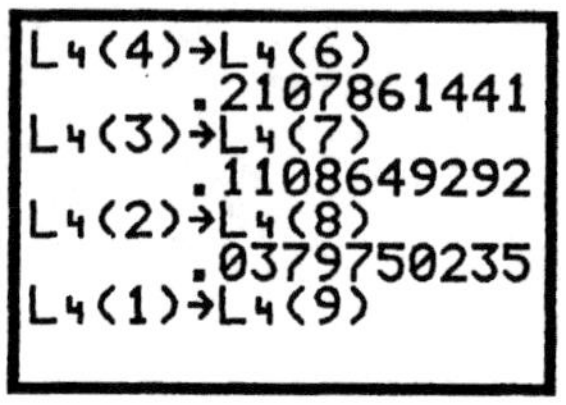

```
L4(4)→L4(6)
       .2107861441
L4(3)→L4(7)
       .1108649292
L4(2)→L4(8)
       .0379750235
L4(1)→L4(9)
```

Now observe the values and compare the "goodness of fit" between the actual frequencies in **L3** versus the expected frequencies in **L4**.

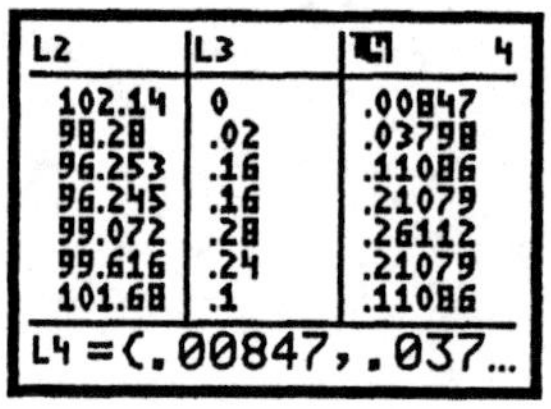

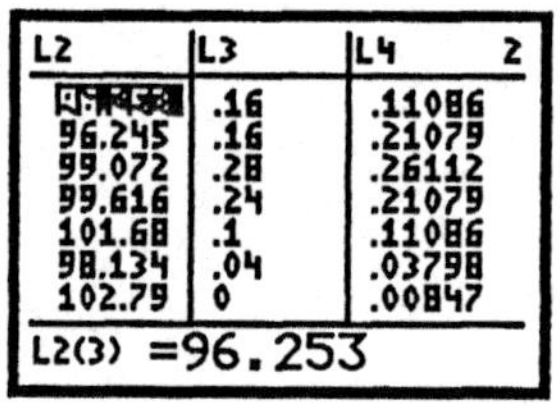

L2	L3	L4	4
102.14	0	.00847	
98.28	.02	.03798	
96.253	.16	.11086	
96.245	.16	.21079	
99.072	.28	.26112	
99.616	.24	.21079	
101.68	.1	.11086	

L4 =(.00847,.037...

L2	L3	L4	2
96.253	.16	.11086	
96.245	.16	.21079	
99.072	.28	.26112	
99.616	.24	.21079	
101.68	.1	.11086	
98.134	.04	.03798	
102.79	0	.00847	

L2(3) =96.253

CHAPTER
4

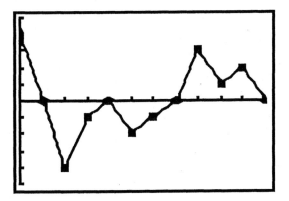

Probability and Sampling Distributions

In this chapter we discuss how to use the TI-83 to generate various types of random numbers. We also provide a supplementary discussion on the geometric distribution along with a program for computing geometric probabilities. Last, we compute probabilities involving the sample mean $\bar{x}$.

4.1 Randomness

We begin by using the TI-83 to generate random numbers. We can generate random numbers between 0 and 1 or between any two values a and b. We can generate random integers between any two integers j and k, and we can also generate random numbers from a normal distribution.

To find the random number generator commands, press **MATH**, then scroll right to **PRB**.

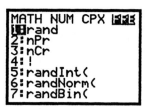

The rand Command

To generate a random number between 0 and 1, enter the **rand** command from the **MATH PRB** menu.

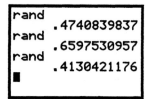

We can also store a sequence of such random numbers into a list. To store ten random numbers between 0 and 1 into list **L2**, enter the command **rand(10)** ➧**L2**.

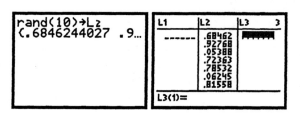

Random Numbers in an Interval

Suppose we wish to generate random numbers that are uniformly distributed throughout an interval $[a, b]$. To do so, we use the command **a+(b−a)rand**.

For example, to choose a random number in the interval [12, 20], enter **12+8rand**.

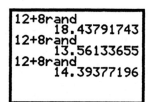

To enter six random numbers from [12, 20] into list **L3**, enter the command **12+8rand(6) ➤L3.**

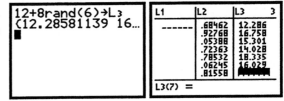

The **SAMPLEN** and **SAMPLEN2** programs from Chapter 3 can be modified easily to generate random numbers from an interval [a, b], rather than from a $N(\mu, \sigma)$ distribution.

The randInt(Command

The command **randInt(j, k)** generates a random integer from j to k. The command **randInt(j, k, n)** generates n such random integers. To the right is a random integer chosen from 10 to 20, and a list of four such random integers.

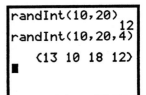

To simulate five rolls of two dice and then to store the sum in list **L3**, enter the commands **randInt(1,6,5) ➤L1**, **randInt(1,6,5)➤L2, L1 + L2➤L3.**

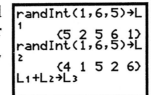

The randNorm(Command

The command **randNorm(μ, σ)** generates a random number from the $N(\mu, \sigma)$ distribution. To the right are random numbers from the $N(68, 4)$ distribution.

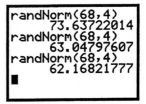

To store ten random numbers from the $N(100, 15)$, distribution in list **L1**, enter the command **randNorm(100, 15, 10) ➤L1.**

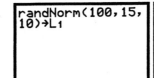

The **randNorm** command was used in the **SAMPLEN** and **SAMPLEN2** programs in Chapter 3.

4.2 Probability Models

In this section we offer a supplementary discussion on the geometric distribution that is used for computing probabilities involving the first time for an event to occur.

The Geometric Distribution

Suppose one has probability p of succeeding on any single independent attempt. Then the geometric distribution counts the *number of attempts needed to obtain the first success.*

We often let $q = 1 - p$ be the probability of failure on any single attempt. Then the **probability (pdf)** of having the first success on the kth attempt, for $k \geq 1$, is given by $q^{k-1}p$. This formula comes from having $k - 1$ failures followed by a success.

The **average number** of attempts needed to succeed is given by $\mu = 1/p$.

The **cumulative probability function (cdf)** gives the probability of succeeding *within* k attempts. Since this event is the complement of failing k times in a row, its probability is given by $1 - q^k$.

The **median** is the smallest integer k for which the cdf equals or exceeds .50. If the median is the integer k, then around half the time we can succeed within k attempts, and around half the time it will take more than k attempts to succeed.

We will also want to find the probability that the first success occurs within j to k attempts. To find this and other probabilities involving a geometric distribution, we can use the built-in **geometpdf(** and **geometcdf(** commands from the bottom of the **DISTR** menu, or we can use a program that prompts the variables.

Rolling a 7 or an 11.   Two dice are rolled over and over until a sum of 7 or 11 is rolled. What is the probability that it takes (a) exactly four rolls? (b) at most six rolls? (c) at least four rolls? (d) from three to six rolls?

Solution: We must first determine the probability p of succeeding on any one attempt. Out of 36 possible outcomes of rolling two dice, there are six ways to roll a 7 and two ways to roll an 11. Thus the probability of rolling a 7 or 11 on any one roll is $p = 8/36 = 2/9$.

(a) To compute the pdf for having the first success on the kth attempt, enter the command **geometpdf(p, k)**. Here use **geometpdf(2/9, 4)** to find the probability that it takes exactly four rolls.

```
geometpdf(2/9,4)
        .1045572321
```

(b) To compute the cdf for succeeding for the first time within k attempts, enter the command **geometcdf(p,k)**. Here use **geometcdf(2/9, 6)** to find the probability that it takes at most six rolls.

```
geometcdf(2/9,6)
        .7786226505
■
```

(c) The event "at least four rolls" is the complement of the event "at most three rolls." So to find the probability that it takes at least four rolls to succeed, enter the command **1 - geometcdf(2/9, 3)**.

```
1-geometcdf(2/9,
3)
        .4705075446
```

(d) The event "three to six" is a difference of events: {1, 2, 3, 4, 5, 6} − {1, 2}. Thus we compute the cdf at 6 and subtract the cdf at 2 to give the probability that it takes from three to six rolls to roll a 7 or 11.

```
geometcdf(2/9,6)
-geometcdf(2/9,2
)
        .3835609221
```

The GEOMET Program

The **GEOMET** program allows the computation of pdf and cdf values by entering the value of p and the lower and upper bounds of j and k. If the same value is entered for both j and k, then it computes the pdf of this value. The program displays the probability along with the median and the average number of attempts needed.

You are also asked if you want a complete distribution to be stored in the **STAT Edit** screen. If so enter **1**; if not enter **0**.

Most of the distribution will be entered into the **STAT Edit** screen if you enter **1**. Under **L1**, the possible numbers of attempts $1, \ldots, n$ are listed. Under **L2**, the pdf values of k, for $1 \le k \le n$, are listed. Under **L3**, the cdf values at k are given. The upper bound n is chosen so that the cdf at n is around .975.

```
PROGRAM:GEOMET
:Disp "PROBABILITY"
:Input P
:Disp "LOWER BOUND"
:Input J
:Disp "UPPER BOUND"
:Input K
:(1-P)^(J-1)-(1-P)^K→C
:int(ln(.5)/ln(1-P))→L
:If ln(.5)/ln(1-P)=L
:Then
:L→L
:Else
:L+1→L
:End
:Disp "COMPLETE DIST?"
:Input Z
:If Z=1
:Then
:ClrList L₁,L₂,L₃
:1→M
:M→L₁(1)
:geometpdf(P,M)→L₂(1)
:geometcdf(P,M)→L₃(1)
:While L₃(M)<.975 and M≤998
:M+1→M
:M→L₁(M)
:geometpdf(P,M)→L₂(M)
:geometcdf(P,M)→L₃(M)
:End
:End
:Disp "PROB=",C
:Disp "MEDIAN=",L
:Disp "AVERAGE=",1/P
```

Exercise. Rework the exercise "Rolling a 7 or 11" using the **GEOMET** program.

Solution: (a) After calling up the **GEOMET** program, enter **2/9** for **PROBABILITY** and enter **4** for both **LOWER BOUND** and **UPPER BOUND**. Also enter **1** to obtain a complete distribution.

The probability of needing exactly four rolls is .1046. The median number of rolls to get a 7 or 11 is found to be 3, while the average number of rolls needed is 4.5.

```
PROBABILITY
?2/9
LOWER BOUND
?4
UPPER BOUND
?4
COMPLETE DIST?
?1
```

```
PROB=
           .1045572321
MEDIAN=
                     3
AVERAGE=
                   4.5
                  Done
```

Since we asked for a complete distribution, observe the lists in the **STAT Edit** screen. For a value of k under **L1**, its pdf is adjacent in **L2** and its cdf is adjacent in **L3**. So for part (b), the probability that it takes at most six rolls is the value in **L3(6)** of .77862.

L1	L2	L3	3
1	.22222	.22222	
2	.17284	.39506	
3	.13443	.52949	
4	.10456	.63405	
5	.08132	.71537	
6	.06325	.77862	
7	.04919	.82782	

L3(6) =.778622650...

For the median value of 3, we see by **L3(3)** that actually 52.95% of the time we should be able to roll a 7 or 11 within three rolls.

Part (c) is worked as before: $1 - \text{cdf}(3) = 1 - .52949 = .47051$.

For part (d), we reexecute the program but enter **3** for **LOWER BOUND** and **6** for **UPPER BOUND**.

```
PROB=
            .3835609221
MEDIAN=
                       3
AVERAGE=
                     4.5
                    Done
```

Betting Red on Roulette. A person playing roulette will continue betting on red until winning. What is the probability of (a) winning within four bets? (b) needing at least three bets to win?

Solution: First note that the chance of winning on a red bet is $p = 18/38$.

For part (a), winning within four bets is the cdf value at 4, meaning that it will take from one to four bets. Either enter the command **geometcdf(18/38, 4)**, or execute the **GEOMET** program by entering **18/38** for **PROBABILITY**, **1** for **LOWER BOUND**, and **4** for **UPPER BOUND**. Either way, we see that there is about a 92.33% chance of winning within four bets.

(b) Enter **1 − geometcdf(18/38, 2)** to see that there is a 27.7% chance of needing at least three bets.

4.3 Sampling Distributions

We now demonstrate how to compute various probabilities involving the sample mean $\bar{x}$.

Sampling from a Normal Distribution

In Section 3.2, we used the **SAMPLEN2** program to simulate the sampling of several random samples size n from a $N(\mu, \sigma)$ distribution. When acquiring such samples, the sample mean $\bar{x}$ follows a $N(\mu, \sigma/\sqrt{n})$ distribution. We can therefore compute the probabilities that a random sample mean will fall between two values, that it will fall higher than a specified value, or that it will fall lower than a specified value.

4.43 ACT scores. During a recent year, ACT scores followed a $N(18.6, 5.9)$ distribution. (a) What is the probability that a single student chosen at random scores 21 or higher? (b) For an SRS of size 50, what are the mean and standard deviation of $\bar{x}$? (c) Compute $P(\bar{x} \geq 21)$.

Solution: (a) We use the **NORMDIST** program to compute $P(X \geq 21)$ for $X \sim N(18.6, 5.9)$. We obtain a value of .3421.

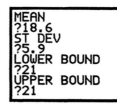 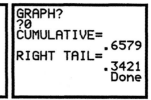

We could also use the built-in command **normalcdf(21, 1E99, 18.6, 5.9)**.

(b) For an SRS of size 50, $\bar{x} \sim N(18.6, 5.9/\sqrt{50}) \approx N(18.6, .8344)$.

(c) The values of $\bar{x}$ should be consistently closer to the mean of 18.9; thus, the probability of $\bar{x}$ being at least 21 should decline.

We verify by computing $P(\bar{x} \geq 21)$ for $\bar{x} \sim N(18.6, 5.9/\sqrt{50})$ with either the **NORMDIST** program or with the command **normalcdf(21, 1E99, 18.6, 5.9/√(50))**. We obtain a value of .002.

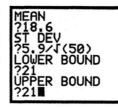 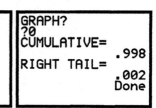

Applying the Central Limit Theorem

When sampling from an arbitrary population with "large" sample sizes n, the sample mean $\bar{x}$ follows an *approximate* $N(\mu, \sigma/\sqrt{n})$ distribution.

4.53 Auto accidents. The number of accidents per week at an intersection varies with mean 2.2 and standard deviation 1.4. Let $\bar{x}$ be the sample mean over the course of 52 weeks. (a) What is the approximate distribution of $\bar{x}$? (b) Approximate $P(\bar{x} < 2)$.

Solution: (a) For 52 weeks, $\bar{x} \approx N(2.2, 1.4/\sqrt{52}) \approx N(2.2, .194145)$. (b) We now compute $P(\bar{x} < 2)$ using either the **NORMDIST** program or the built-in **normalcdf(** command. Using the built-in command, we note that $P(\bar{x} < 2) = P(-1E99 \leq \bar{x} < 2)$. Either way, we see that $P(\bar{x} < 2) \approx .1515$.

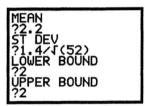 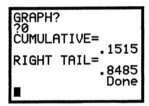 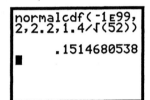

4.54 Pollutants in auto accidents. The level of NOX in the exhaust of a car model varies with mean 0.9 g/mi and standard deviation 0.15 g/mi. For a company's fleet of 125 such cars, (a) what is the approximate distribution of $\bar{x}$? (b) What is the level L such that the probability of $\bar{x}$ being greater than L is only 0.01?

Solution: (a) For 125 cars, $\bar{x} \approx N(0.9, 0.15/\sqrt{125}) \approx N(0.9, .0134)$. (b) We now wish to find the inverse cumulative normal value L for which $P(\bar{x} > L) = .01$, or equivalently $P(\bar{x} \leq L) = .99$. We compute this value with the **invNorm(** command from the **DISTR** menu.

We see that only 1% of the time should $\bar{x}$ be larger than $L = .9312$.

CHAPTER
5

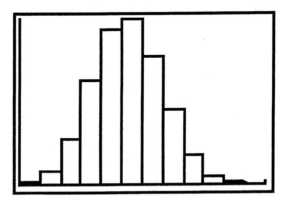

Probability
Theory

5.1 General Probability Rules
5.2 The Binomial Distributions
5.3 Conditional Probability

In this chapter we discuss how to use the TI-83 to compute binomial probabilities and provide a program for doing so. We also provide supplementary programs for the hypergeometric distribution and for the Law of Total Probability and Bayes' Rule.

5.1 General Probability Rules

We begin by providing a supplementary discussion on the hypergeometric distribution that models sampling without replacement. The material in this section is optional and is not needed for further sections.

The Hypergeometric Distribution

Suppose we have a population of N objects that are divided into two types: Type A and Type B. There are r objects of Type A and $N - r$ objects of Type B. For example, a standard deck of 52 playing cards can be divided in many ways. Type A could be "hearts" and Type B could be "all others." In this case there are 13 hearts and 39 others.

Suppose a random sample of size n is taken all at once (without replacement) from the entire population of N objects. The hypergeometric distribution X counts the total number of objects of Type A in the sample.

The probability of having exactly k objects of Type A is given by

$$P(X = k) = \frac{\binom{r}{k}\binom{N-r}{n-k}}{\binom{N}{n}}.$$

The *average number* of objects of Type A in the sample is given by

$$\mu = n\left(\frac{r}{N}\right).$$

The *variance* in the number of objects of Type A in the sample is given by

$$\sigma^2 = n\left(\frac{r}{N}\right)\left(1 - \frac{r}{N}\right)\left(\frac{N-n}{N-1}\right).$$

The HYPGEOM Program

There is no known formula for the cumulative probability $P(X \leq k)$ or for computing probabilities such as $P(j \leq X \leq k)$. Following is a program for computing such probabilities.

```
PROGRAM:HYPGEOM
:Disp "POP. SIZE"
:Input N
:Disp "TYPE A SIZE"
:Input M
:Disp "SAMPLE SIZE"
:Input R
:Disp "LOWER BOUND"
:Input J
:Disp "UPPER BOUND"
:Input K
:sum(seq(M nCr I*(N-M) nCr (R- I)/N nCr R,I,J,K,1))→C
:Disp "COMPLETE DIST?"
:Input Z
:If Z=1
:Then
:max(0,R-(N-M))→L
:seq(I,I,L,min({M,R,998}))→L₁
:seq(M nCr I*(N-M) nCr (R-I)/N nCr R,I,L,min({M,R,998}))→L₂
:cumSum(L₂)→L₃
:End
:Disp "PROB="
:Disp C
:Disp "AVE="
:Disp R*M/N
:Disp "ST DEV="
:Disp √(R*M/N*(N-M)/N*(N-R)/(N-1))
```

To execute the program, we enter the values of the population size N, the Type A size r, the sample size n, and the lower and upper bounds for computing the probability $P(j \leq X \leq k)$. To compute $P(X = k)$, enter the same value k for both the lower bound and the upper bound.

If you would like a complete distribution entered into the **STAT Edit** screen, then enter **1** for **COMPLETE DIST?** Otherwise, enter **0**. If you enter **1**, then under **L1** will be listed the possible numbers of objects from Type A. Under **L2**, the values of $P(X = k)$ will be listed. Under **L3**, the values of $P(X \leq k)$ will be listed.

The program displays $P(j \leq X \leq k)$, along with the average value and standard deviation of the distribution.

The chance of hearts. Suppose a hand of five cards is dealt. What is the probability of there being (a) at least three hearts? (b) exactly three hearts? (c) What is the average number of hearts and the most likely number of hearts to be dealt?

Solution: There are 13 hearts in a deck of 52. In this sample of 5, "at most three" means from 3 to 5. After calling up the **HYPGEOM** program, enter **52** for **POP. SIZE**, enter **13** for **TYPE A SIZE**, enter **5** for **SAMPLE SIZE**, enter **3** for **LOWER   BOUND**, and enter **5** for **UPPER   BOUND**. Also enter **1** for **COMPLETE DIST?**

Observe the display, then view the lists in the **STAT EDIT** screen.

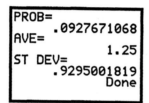

We see that $P(3 \le X \le 5) \approx .0928$. There would be an average of 1.25 hearts in a random 5-card deal with a standard deviation of 0.9295. From list **L2**, we see that the chance of obtaining exactly three hearts is $P(X = 3) = .08154$. Also from list **L2**, we see that we would most likely have 1 heart and $P(X = 1) = .41142$.

How many men? In a drama class, there are 12 women and 10 men. A group of 15 students is to be chosen at random to read a screenplay. What is the probability that there will be at most six men chosen? At least six men? What is the average number of men and the most likely number of men to be chosen?

Solution: There are 10 men (Type A). There are 15 people chosen and there are only 12 women; thus, there must be at least 3 men chosen. So in this case, "at most 6" means 3 to 6 men. In the program, enter **22** for **POP. SIZE**, enter **10** for **TYPE A SIZE**, enter **15** for **SAMPLE SIZE**, enter **3** for **LOWER BOUND**, and enter **6** for **UPPER BOUND**. Also enter **1** for **COMPLETE DIST?**

```
PROB=
        .3839009288
AVE=
        6.818181818
ST DEV=
        1.113404429
                Done
```

L1	L2	▮▮ 3
3	7E-4	7E-4
4	.01478	.01548
5	.09752	.113
6	.2709	.3839
7	.3483	.7322
8	.20898	.94118
9	.05418	.99536

L3 =(7.036307346...

We see that $P(3 \leq X \leq 6) = .3839$ and the average number of males chosen in a sample of 15 will be 6.8182. Under **L2** we find the mode to be 7, which occurs with probability .3483.

Since there are 10 males, "at least six" means from 6 to 10. If we rerun the program with **6** for **LOWER BOUND** and **10** for **UPPER BOUND**, we will find that $P(6 \leq X \leq 10) = .887$.

How many women? In another class, there are 12 men and 17 women. A random group of 14 is chosen. What is the probability of there being at most nine women chosen? At least nine women? What is the most likely number of women to be chosen?

Solution: There are 17 women (Type A). There are 14 chosen and there are only 12 men; thus, there must be at least 2 women chosen. So in this case, "at most nine" means 2 to 9 women. In the program, enter **29** for **POP. SIZE**, enter **17** for **TYPE A SIZE**, enter **14** for **SAMPLE SIZE**, enter **2** for **LOWER BOUND**, and enter **9** for **UPPER BOUND**. Also enter **1** for **COMPLETE DIST?**

```
PROB=
        .8351298035
AVE=
        8.206896552
ST DEV=
        1.348800498
                Done
```

L1	L2	L3	3
2	1.8E-6	1.8E-6	
3	1.1E-4	1.1E-4	
4	.00203	.00213	
5	.01755	.01968	
6	.07899	.09867	
7	.1986	.29727	
8	.28962	.58689	

L3(1)=1.753509210...

L1	L2	L3	3
9	.24824	.83513	
10	.12412	.95925	
11	.03511	.99436	
12	.00527	.99962	
13	3.7E-4	.99999	
14	8.8E-6		

L3(14) =

We see that $P(2 \leq X \leq 9) \approx .8351$. Under **L2**, we find the mode to be 8, which occurs with probability .28962.

Here "at least nine" means from 9 to 14. If we rerun the program with **9** for **LOWER BOUND** and **14** for **UPPER BOUND**, we will see that $P(9 \leq X \leq 14) \approx .41311$.

5.2 The Binomial Distributions

We now demonstrate how to compute various probabilities from a given binomial distribution, including the normal approximation of the binomial.

Computing Binomial Probabilities

For a binomial distribution $X \sim b(n,p)$, we can compute probabilities using the built-in **binompdf(** and **binomcdf(** commands (items **0** and **A**) from the **DISTR** menu. We compute the probability $P(X = k)$ of exactly k occurrences by entering the command **binompdf(n, p, k)**. The probability $P(X \le k) = P(0 \le X \le k)$ of at most k occurrences is computed with the command **binomcdf(n, p, k)**.

For bounds j and k, such that $1 \le j < k \le n$, the probability of there being from j to k occurrences is given by $P(j \le X \le k) = P(X \le k) - P(X \le j-1)$ = **binomcdf(n, p, k) - binomcdf(n, p, j–1)**.

The mean number of occurrences is always given by $\mu = np$.

Exercise 5.22.   A factory employs several thousand workers of whom 30% are Hispanic. If the 15 union executive committee members were chosen from the workers at random, then the number of Hispanic workers on the committee would have the binomial distribution with $n = 15$ and $p = 0.3$.
(a) What is the probability that exactly 3 committee members are Hispanic?
(b) What is the probability that 3 or fewer committee members are Hispanic?

Solution: Letting $X \sim b(15,0.30)$, we must find (a) $P(X = 3)$ and (b) $P(X \le 3)$.

For $P(X = 3)$ we enter **binompdf(15, .3, 3)**. For $P(X \le 3)$ we enter **binomcdf(15, .3, 3)**.

```
binompdf(15,.3,3
)
         .1700402133
binomcdf(15,.3,3
)
         .2968679279
```

The BINOMIAL Program

We can also compute various probabilities with a program that prompts the variables.

```
PROGRAM:BINOMIAL
:Disp "NUMBER OF TRIALS"          :0→Xmin
:Input N                          :N→Xmax
:Disp "PROBABILITY"               :1→Xscl
:Input P                          :0→Ymin
:Disp "LOWER BOUND"               :max(L₂)→Ymax
:Input J                          :1→Yscl
:Disp "UPPER BOUND"               :PlotsOff
:Input K                          :PlotsOn 1
:sum(seq(binompdf(N,P,I),I,J,K))→C :Plot1(Histogram,L₁,L₂)
:Disp "COMPLETE DIST?"            :End
:Input Z                          :Disp "PROB="
:If Z=1                           :Disp C
:Then                             :Disp "MEAN="
:FnOff                            :Disp NP
:seq(I,I,0,N)→L₁                  :Disp "ST DEV="
:binompdf(N,P)→L₂                 :Disp √(NP(1-P))
:binomcdf(N,P)→L₃
```

To execute the **BINOMIAL** program, we enter the value of n for **NUMBER OF TRIALS**, the value of p for **PROBABILITY**, and the values of the bounds j and k for computing the probability $P(j \le X \le k)$. To compute $P(X = k)$, enter the same value k for both the lower bound and the upper bound.

The program also asks if you would like the complete distribution to be entered into the **STAT EDIT** screen. If so, enter **1**; if not, enter **0**. The program displays $P(j \le X \le k)$, along with the mean and standard deviation of the distribution.

If the complete distribution is entered, then list **L1** will contain the possible numbers of occurrences $0, 1, \ldots, n$. Under **L2**, the values of $P(X = k)$, for $0 \le k \le n$, will be listed. Under **L3**, the values of $P(X \le k)$, for $0 \le k \le n$, will be listed. We note that a complete distribution can be entered only for $n \le 998$.

Exercise. Rework Exercise 5.22 with the **BINOMIAL** program.

Solution: We enter the parameters into the **BINOMIAL** program. To find $P(X=3)$, we enter **3** for both **LOWER BOUND** and **UPPER BOUND**. We also enter **1** for a complete distribution. The value of $P(X=3)$ is displayed, along with the mean and standard deviation.   By looking at the value adjacent to 3 in list **L2**, we verify that $P(X=3) \approx .17$. By looking at the value adjacent to 3 in list **L3**, we see that $P(X \leq 3) \approx .29687$.   (Alternately, we could rerun the program with bounds of 0 and 3.)

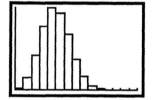

As an extra feature, press **GRAPH** to see a probability histogram.

5.36 Lie detectors. A federal report finds that lie detector tests given to truthful persons have a probability of about 0.2 of suggesting that the person is deceptive.
(a)   A company uses a lie detector on 12 job applicants to assess their truthfulness about previous employers. Suppose all 12 answer truthfully. What is the probability that the lie detector says that all 12 are truthful? What is the probability that the lie detector says that at least one person is deceptive?
(b) What are the mean and standard deviation of the number of 12 truthful persons who will be classified as deceptive?
(c) What is the probability that the number classified as deceptive is less than the mean?

Solution: (a) We shall let X denote the number of those classified as deceptive. Then $X \sim b(12, 0.20)$. The probability of all 12 being classified as truthful is then $P(X=0)$. The probability that at least one is classified as deceptive is $P(X \geq 1) = 1 - P(X=0)$.

These probabilities can be easily calculated using the binomial formula

$$P(X = 0) = (1 - p)^n = (.80)^{12} \approx 0.06872.$$

Thus, $P(X \geq 1) \approx 1 - 0.06872 = 0.93128.$

We can verify the result with the **BINOMIAL** program or the built-in **binompdf(** command.

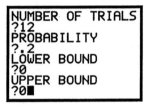

Using the program, we also obtain the answers to (b). For part (c), we wish to find $P(0 \leq X \leq 2)$. If we have entered **1** for a complete distribution while running the **BINOMIAL** program in part (a), then we can use list **L3** to see that $P(0 \leq X \leq 2) \approx .55835$. Otherwise, we could reexecute the program with **0** and **2** as the bounds or we could use the built-in **binomcdf(** command.

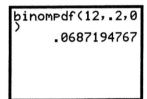

Normal Approximation of Binomial

5.29 Checking for survey errors. About 12% of American adults are black. Therefore the number of blacks in a random sample of 1500 adults should vary with the binomial distribution $X \sim b(1500, 0.12)$. (a) What are the mean and standard deviation of X? (b) Use the normal approximation to find $P(X \leq 170)$.

Solution: (a) The mean is $\mu = 1500(.12) = 180$, and the standard deviation is $\sigma = \sqrt{1500(.12)(.88)} \approx 12.5857.$

(b) We now let $Y \sim N(\mu, \sigma) = N(180, 12.5857)$. Then $P(X \le 170) \approx P(Y \le 170)$. We can compute this left-tail normal probability using either the **NORMDIST** program or the built-in **normalcdf(** command. Using the program, we recall that we enter the same endpoint 170 for both **LOWER BOUND** and **UPPER BOUND** to find tail values. Using the built-in command, we recall that $P(Y \le 170) \approx P(-1E99 \le Y \le 170)$.

We see that $P(X \le 170) \approx .2134$.

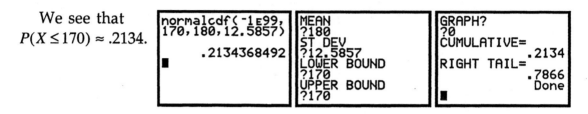

Note: With the value of $n = 1500$ in $X \sim b(1500, 0.12)$, the TI-83 overflows when attempting to use either the **binomcdf(** command or the **BINOMIAL** program to compute $P(X \le 170)$.

5.3 Conditional Probability

We conclude this chapter with a supplemental program for executing the Law of Total Probability and Bayes' Rule.

Total Probability

Suppose we wish to find the probability of event A. Assume we know conditional probabilities $P(A|B_1)$, $P(A|B_2)$, ..., $P(A|B_n)$ for other events B_1, B_2, ..., B_n that form a partition of the population. Assume that we also know $P(B_1)$, ..., $P(B_n)$. Then the total probability of event A is

$$P(A) = P(B_1)P(A|B_1) + P(B_2)P(A|B_2) + ... + P(B_n)P(A|B_n).$$

The probabilities of the intersections $P(A \cap B_1)$, ..., $P(A \cap B_n)$ are actually the individual summands in the formula for $P(A)$. That is, $P(A \cap B_i) = P(B_i)P(A|B_i)$, for $i = 1, ... n$.

After we find $P(A)$, then we can apply Bayes' Rule to find the reverse conditional probabilities $P(B_1|A)$, ..., $P(B_n|A)$.

The BAYES Program

We can compute the total probability, as well as various other conditional probabilities, with the **BAYES** program. Before executing the program, enter the known probabilities $P(B_1)$, . . . , $P(B_n)$ into list **L1** and the known conditionals $P(A|B_1)$, . . . , $P(A|B_n)$ into list **L2**. The program displays the total probability $P(A)$, stores the probabilities of the intersections $P(A \cap B_1)$, . . . , $P(A \cap B_n)$ in list **L3**, and stores the reverse conditionals $P(B_1|A)$, . . . , $P(B_n|A)$ in list **L4**.

The program also computes and stores two other lists of conditionals. The conditional probabilities $P(B_1|A')$, . . . , $P(B_n|A')$ are stored in list **L5**, and the conditional probabilities $P(A|B_1')$, . . . , $P(A|B_n')$ are stored in list **L6**.

To help the user keep track of which list contains which probabilities, the program displays a description.

```
PROGRAM:BAYES
:sum(seq(L₁(I)L₂(I),I,1,dim(L₁)))→T
:L₁*L₂→L₃
:L₃/T→L₄
:L₁*(1-L₂)/(1-T)→L₅
:T(1-L₄)/(1-L₁)→L₆
:Disp "TOTAL PROB"
:Disp round(T,4)
:Disp "A AND Bs : L₃"
:Disp "Bs GIVEN A : L₄"
:Disp "Bs GIVEN A' : L₅"
:Disp "A GIVEN B's : L₆"
```

Counting the vote. Among voters in a large city, 40% are registered Democrats, 35% are registered Republicans, and 25% are registered as Independents or with another party. In a close election, the incumbent mayor expects to receive 30% of the Democrat vote, 80% of the Republican vote, and 40% of the remaining vote. Apply the **BAYES** program to compute the percent of the overall vote that the candidate expects to receive. Then analyze the other computed conditional probabilities.

Solution: Here we let B_1 = registered Democrats, B_2 = registered Republicans, and B_3 = other registered voters. We let A be the event that a person votes for the candidate. Then $P(A|B_1)$ = .30, $P(A|B_2)$ = .80, and $P(A|B_3)$ = .40. We can now enter the given probabilities and conditionals into **L1** and **L2** and then execute the program.

We see that the candidate can expect to receive 50% of the overall vote. The other computed probabilities are now stored in the designated lists.

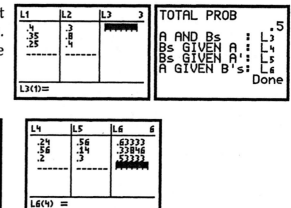

From **L3**, the probability that one is a Democrat and will vote for the candidate is $P(A \cap B_1)$ = .12. The probability that one is a Republican and will vote for the candidate is $P(A \cap B_2)$ = .28. The probability that one is Independent/third party and will vote for the candidate is $P(A \cap B_3)$ = .1.

From **L4**, given that one *will* vote for the candidate, the respective conditional probabilities of being Democrat, Republican, or Independent/third party are $P(B_1|A)$ = .24, $P(B_2|A)$ = .56, and $P(B_3|A)$ = .20. (These values must sum to 1.) This list is the direct application of Bayes' Rule.

From **L5**, given that one will *not* vote for the candidate, the respective conditional probabilities of being Democrat, Republican, or Independent/third party are $P(B_1|A')$ = .56, $P(B_2|A')$ = .14, and $P(B_3|A')$ = .30. (Again these values must sum to 1.)

From **L6**, given that a voter is not Democrat, the conditional probability of voting for the candidate is $P(A|B_1')$ = .63333. Given that a voter is not Republican, the conditional probability of voting for the candidate is $P(A|B_2')$ = .33846. Given that a voter is not Independent/third party, the conditional probability of voting for the candidate is $P(A|B_3')$ = .53333.

We note that given any conditional probability $P(C|D)$, the complement conditional probability is given by $P(C'|D) = 1 - P(C|D)$. Thus, lists for complement conditional probabilities do not need to be generated. For example, the respective conditional probabilities of *not* voting for the candidate given that one is Democrat, Republican, or Independent/third party are .70, .20, .60. These values are the complement probabilities of the originally given conditionals.

CHAPTER
6

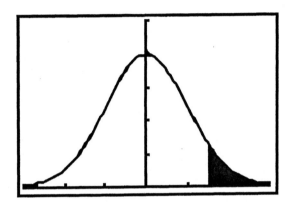

Introduction
to Inference

6.1 Estimating with Confidence
6.2 Tests of Significance
6.3 Making Sense of Statistical Significance
6.4 Error Probabilities and Power

In this chapter, we discuss how to use the TI-83 to compute confidence intervals and conduct hypothesis tests for the mean μ of a normally distributed population with known standard deviation σ.

6.1 Estimating with Confidence

In this section, we discuss how to find a confidence interval for the mean of a normal population with known standard deviation σ. To access the required TI-83 command, press **STAT**, scroll right to **TESTS**, then press **7** for **ZInterval**.

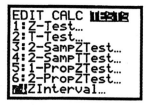

6.4 Surveying hotel managers.   In a survey of 160 hotel general managers, there were 114 responses. The average time these 114 general managers had spent with their current company was 11.78 years. It is known that the standard deviation of time for all general managers is 3.2 years. Give a 99% confidence interval for the mean number of years hotel general managers have spent with their current company.

Solution: Press **STAT**, scroll right to **TESTS**, and press **7** for **ZInterval**. Set the **Inpt** by highlighting **Stats** and pressing **ENTER**. Next enter the values of σ = 3.2, $\bar{x}$ = 11.78, n = 114, and C-Level = .99. Then scroll down to **Calculate** and press **ENTER**.

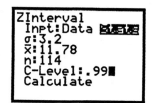 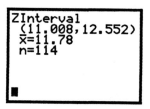

We obtain a 99% confidence interval of (11.008, 12.552).

6.5 IQ test scores. Here are the IQ scores from an SRS of 31 seventh-grade girls in a midwest school district.

114	100	104	89	102	91	114	114	103	105	
108	130	120	132	111	128	118	119	86	72	
111	103	74	112	107	103	98	96	112	112	93

Suppose that the standard deviation of IQ scores for all seventh-grade girls in the school district is known to be 15. Give a 99% confidence interval for the mean IQ score of this population.

Solution: First enter the data into a list, say list **L1**. Then press **STAT**, scroll right to **TESTS**, and press **7** for **ZInterval**. Set the **Inpt** by highlighting **Data** and pressing **ENTER**. Next adjust the value of σ, the list, frequency, and confidence level. Then scroll down to **Calculate** and press **ENTER**.

The calculator will compute the sample mean $\bar{x}$ from the data in the list, then display the confidence interval and $\bar{x}$.

```
ZInterval
Inpt:Data Stats
σ:15
List:L1
Freq:1
C-Level:.99
Calculate
```

```
ZInterval
(98.899,112.78)
x̄=105.8387097
Sx=14.27140912
n=31
```

The built-in **ZInterval** command rounds the confidence intervals to a certain number of decimal places. If one desires more accuracy, then one can find the actual interval "by hand" by finding the critical value $z*$ and then applying the formula

$$\mu \approx \bar{x} \pm \frac{z^* \sigma}{\sqrt{n}}.$$

Finding Critical Values

The standard levels of confidence are .90, .95, .98, and .99. The .90 critical value is the number $z*$ such that $P(-z* \leq Z \leq z*) = .90$, where $Z \sim N(0, 1)$. This critical value is represented as the right endpoint of the shaded area of the $N(0, 1)$ curve below. The shaded area represents .90 probability.

For the .90 critical value, there is .05 probability remaining on each tail. Thus at the .90 critical value, the *cumulative* probability is .95 (from .90 on the inside and .05 at the left tail).

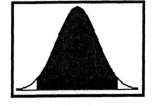

To find this critical value, call up the **invNorm** command from the **DISTR** menu, then enter the command **invNorm(.95, 0, 1)**.

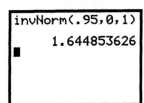

```
invNorm(.95,0,1)
          1.644853626
```

We commonly round the .90 critical value to 1.645.

Other critical values. Find the .95 and .99 critical values.

Solution: For the .95 critical value, there is .025 at each tail; thus the cumulative probability is .975 at the critical value. For the .99 critical value, there is .005 at each tail; thus the cumulative probability is .995 at the critical value.

```
invNorm(.975,0,1
)
          1.959963986
invNorm(.995,0,1
)
          2.575829303
```

We commonly round the .95 critical value to 1.96 and the .99 critical value to 2.576.

6.7 Confidence level and interval length. A sample of three measurements gave a value of $\bar{x} = 0.8404$. It is known that $\sigma = 0.0068$. (a) Find an 80% confidence interval for μ. (b) Find a 99.9% confidence interval for μ.

Solution: We will find the interval "by hand" and then compare the results with the built-in **ZInterval** command. (a) We first find the .80 critical value. There is .10 probability remaining at each tail, so there is .90 cumulative probability at the critical value.

We compute the critical value and store it to the variable **Z** on the calculator. Then we type in the endpoints of the confidence interval.

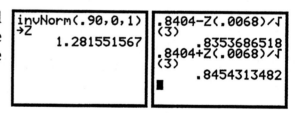

We see that the interval to six decimal places is (.835369, .845431).

(b) We now find the .999 critical value. There is .0005 probability remaining at each tail; so there is .9995 cumulative probability at the critical value.

We again compute the critical value and store it to the variable **Z** on the calculator. Then we type in the endpoints of the confidence interval.

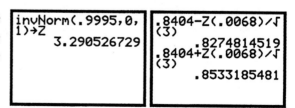

The interval to six decimal places is (.827481, .853319). As shown next, both of these intervals can be verified with the **ZInterval** command.

ZInterval
 Inpt:Data **STATS**
 σ:.0068
 x̄:.8404
 n:3
 C-Level:.8
 Calculate

ZInterval
 (.83537,.84543)
 x̄=.8404
 n=3

ZInterval
 Inpt:Data **STATS**
 σ:.0068
 x̄:.8404
 n:3
 C-Level:.999
 Calculate

ZInterval
 (.82748,.85332)
 x̄=.8404
 n=3
 ■

Choosing the Sample Size

To find the sample size that will give a specified level of confidence and margin of error, we apply the formula

$$ n \geq \left(\frac{z^* \sigma}{m} \right)^2, $$

where z^* is the computed critical value and m is the desired margin of error.

6.10 Calibrating a scale. The scale readings of a 10-gram weight are normally distributed with a known standard deviation of 0.0002 grams. (a) If the weight is weighed five times with a mean of 10.0023 grams, give a 98% confidence interval for the mean reading of the weight. (b) How many measurements are needed to obtain a margin of error of ± 0.0001 with 98% confidence?

Solution: Call up the **ZInterval** command. Set the **Inpt** to **Stats**, enter .0002 for σ, enter 10.0023 for $\bar{x}$, and enter 5 for n. Then adjust the confidence level, scroll down to **Calculate**, and press **ENTER**.

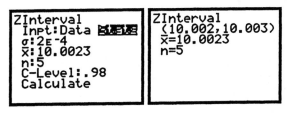

Before we compute the desired sample size in part (b), it is important to observe the round-off problem in this calculation. The sample mean is 10.0023, which should be in the center of the confidence interval. However, the right endpoint has been rounded up to 10.003, which makes the interval much wider than it should be. However, as in Exercise 6.7, we can find the actual interval "by hand" by finding the critical value z^*, which is also needed to find the sample size in (b).

We now find this .98 critical value. There is .01 probability remaining at each tail, so there is .99 cumulative probability at the critical value.

We compute the critical value and store it to the variable **Z** on the calculator. Then we type in the endpoints of the confidence interval.

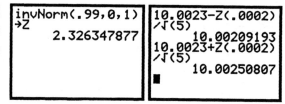

We see that the interval to four decimal places is (10.0021, 10.0025), which is symmetric about the value of $\bar{x} = 10.0023$.

We can now compute the desired sample size for $m = .0001$ and our known value of $\sigma = .0002$. We would need a sample of size 22 to obtain a margin of error of $\pm\ 0.0001$ with 98% confidence.

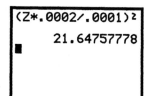

The ZSAMPSZE Program

We could also use a program that computes (rounded-up) sample sizes. To execute the **ZSAMPSZE** program, we input the value of the standard deviation σ, the desired margin of error, and the desired confidence level.

```
PROGRAM:ZSAMPSZE
:Disp "ST DEV"                    :If int(M)=M
:Input S                          :Then
:Disp "DESIRED ERROR"             :M→N
:Input E                          :Else
:Disp "CONF. LEVEL"               :int(M+1)→N
:Input R                          :End
:invNorm((R+1)/2,0,1)→Q           :Disp "SAMPLE SIZE="
:(Q*S/E)²→M                       :Disp int(N)
```

Reworking Exercise 6.10 (b) with the program:

6.2 Tests of Significance

We now discuss how to use the TI-83 to conduct one-sided and two-sided hypothesis tests about the mean μ of a normally distributed population for which the standard deviation σ is known. To access the required command, press **STAT**, scroll right to **TESTS**, then press **1** for **Z-Test**.

6.34 Sales of coffee. Weekly sales of coffee vary normally with a mean of $\mu = 354$ units and a standard deviation of $\sigma = 33$ units. After reducing the price, sales in the next three weeks are 405, 378, and 411 units. Is there evidence to suggest that sales are now higher? Perform the test H_0: $\mu = 354$, H_a: $\mu > 354$. Give the z-test statistic and the P-value. Sketch a curve with the area corresponding to the P-value shaded.

Solution: We can quickly compute the z-test statistic and the P-value with the TI-83's built-in command. First enter the data into a list, say list **L1**. Then call up the **Z-Test** screen from the **STAT TESTS** menu, and adjust the **Inpt** to **Data**.

Enter the values $\mu_0 = 354$ and $\sigma = 33$, then set the list to **L1** and press **E N T E R** on the alternative $>\mu_0$. Then press **ENTER** on **Calculate**.

We obtain a test statistic of 2.3094 and a P-value of .0105. If the true mean were still equal to 354 with $\sigma = 33$, then there would be only a 1.05% chance of the sample group averaging as high as $\bar{x} = 398$. This low P-value means that there is significant evidence to conclude that the mean is now higher than 354.

To see a sketch, reexecute the **Z-Test** command, but press **ENTER** on **Draw**.

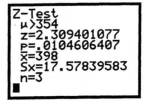

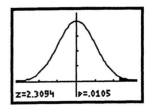

6.35 Engine crankshafts.   Here are the measurements (in millimeters) of a critical dimension on a sample of crankshafts.

224.120	224.001	224.017	223.982	223.989	223.961
223.960	224.089	223.987	223.976	223.902	223.980
224.098	224.057	223.913	223.999		

The manufacturing process varies normally with a standard deviation of $\sigma = 0.060$ mm.   The process mean is supposed to be 224 mm, and we wish to use the data to test if it is so.   State the appropriate H_0 and H_a, then conduct the test and give the statistic and the P-value.

Solution: We shall test whether the mean is 224, H_0: $\mu = 224$, with the alternative hypothesis that it is not 224, H_a: $\mu \neq 224$.

Enter the data into a list, say list **L2**. Then call up the **Z-Test** screen from the **STAT TESTS** menu and adjust the **Inpt** to **Data**.

Enter the values $\mu_0 = 224$ and $\sigma = .06$, then set the list to **L2** and press **ENTER** on the alternative $\neq \mu_0$. Then press **ENTER** on **Calculate**.

We obtain a test statistic of .129 and a P-value of .8972. If the true mean were equal to 224 with $\sigma = 0.060$, then there would be an 89.72% chance of a sample group of 16 averaging as far away from 224 as $\bar{x} = 224.0019$. There is no significant evidence to conclude that the mean is not 224.

6.43 Is this milk watered down? The freezing temperature of milk varies normally with mean $\mu = -0.545\,°C$ and standard deviation $\sigma = 0.008\,°C$. Added water raises the freezing temperature. A laboratory measured the freezing temperature of five lots of milk from one producer and found a sample mean of $\bar{x} = -0.538\,°C$. State and perform a test to see if there is evidence to suggest that the milk is watered down. Give the z-test statistic and the P-value.

Solution: We shall perform the test H_0: $\mu = -0.545$ versus the alternative H_a: $\mu > -0.545$. Call up the **Z-Test** feature from the **STAT TESTS** menu and adjust the **Inpt** to **STATS**, which allows us to enter the variables of our choice. Enter the values $\mu_0 = -.545$, $\sigma = .008$, $\bar{x} = -.538$, and $n = 5$.

Highlight and enter the alternative, in this case $> \mu_0$, then scroll down to **Calculate** and press **ENTER**.

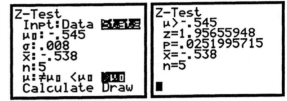

We obtain a z-test statistic of 1.957 and a P-value of .0252. Since the P-value is rather small, we have significant evidence to reject a claim that the average freezing temperature from this producer is $-.545$. If it were, then there would be only a .0252 probability of having a sample mean as high as $\bar{x} = -.538$. We have evidence to suggest that this milk is watered down.

We now provide an example of how to conduct a two-sided test, with level of significance α, based on a $(1-\alpha) \times 100\%$ confidence interval.

6.5 IQ test scores. Here are the IQ scores of an SRS of 31 seventh-grade girls in a midwest school district.

114	100	104	89	102	91	114	114	103	105	
108	130	120	132	111	128	118	119	86	72	
111	103	74	112	107	103	98	96	112	112	93

Suppose that the standard deviation of all seventh-grade girls in the school district is known to be 15. (a) Give a 95% confidence interval for the mean IQ score of this population. (b) Is there significant evidence at the 5% level to conclude that the mean IQ score of the population differs from 100? State hypotheses and base a test on the confidence interval from (a).

Solution: (a) We first enter the data into a list, say **L3**. Then we compute the confidence interval with the **ZInterval** feature from the **STATS TESTS** menu.

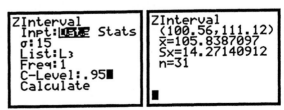

(b) We now test H_0: $\mu = 100$, H_a: $\mu \neq 100$. Since the value of 100 does not fall within the confidence interval, we have significant evidence at the 5% level to reject the hypothesis that $\mu = 100$.

6.3 Making Sense of Statistical Significance

We continue with two other examples that illustrate how one must be careful in drawing conclusions of significance.

6.54 Is it significant? Suppose that SATM scores vary normally with $\mu = 475$ and $\sigma = 100$. One hundred students go through rigorous training before the test to see if there will be an improvement in average score. Consider the test H_0: $\mu = 475$, H_a: $\mu > 475$. (a) Is $\bar{x} = 491.4$ significant at the 5% level? (b) Is $\bar{x} = 491.5$ significant at the 5% level?

Solution: We perform the hypothesis test for both values of $\bar{x}$ using the **Z-Test** feature from the **STAT TESTS** menu.

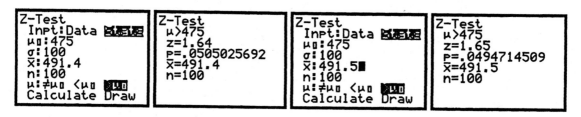

In the first case, $P = .0505 > .05$, so the value of $\bar{x} = 491.4$ is not significant at the 5% level. However in the second case, $P = .04947 < .05$, so the value of $\bar{x} = 491.5$ is significant at the 5% level. However for SATM scores, there is no real "significant" difference between means of 491.4 and 491.5.

6.55 Coaching and the SAT. Suppose again that SATM scores vary normally with $\sigma = 100$. Calculate the P-value for the test of H_0: $\mu = 475$, H_a: $\mu > 475$ in each of the following situations.
(a) A sample of 100 coached students yielded an average of $\bar{x} = 478$.
(b) A sample of 1,000 coached students yielded an average of $\bar{x} = 478$.
(c) A sample of 10,000 coached students yielded an average of $\bar{x} = 478$.

Solution: We again adjust the settings in the **Z-Test** screen from the **STAT TESTS** menu. Following are the results using the three different sample sizes.

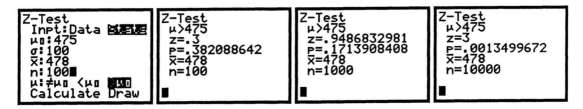

We see that the rise in average to $\bar{x} = 478$ is significant (P very small) only when the results stem from the very large sample of 10,000 coached students. With the sample of only 100 students, there is 38.2% chance of obtaining a sample mean as high as $\bar{x} = 478$ even if the true mean were still 475.

6.4 Error Probabilities and Power

We conclude this chapter with some exercises on computing the power against an alternative and computing the probabilities of Type I and Type II errors.

Exercise 6.67. Consider the hypotheses H_0: $\mu = 300$, H_a: $\mu < 300$ at a 5% level of significance. A sample of size $n = 6$ is taken from a normal population having $\sigma = 3$. (a) Find the power of this test against the alternative $\mu = 299$. (b) Find the power of this test against the alternative $\mu = 295$.

Solution: We first find the rejection region of the test at the 5% level of significance. Since the alternative is the one-sided left tail, we wish the left-tail probability under the standard normal curve to be .05. This probability occurs at $z = -1.645$ (the negative of the .90 critical value). So we reject H_0 if the z-test statistic is less than -1.645. That is, we reject if

$$\frac{\bar{x} - 300}{3 / \sqrt{6}} < -1.645$$

or equivalently if $\bar{x} < 300 - 1.645(3 / \sqrt{6}) = 297.985$. Now we must find the probability that $\bar{x}$ is less than 297.985 given that the alternative $\mu = 299$ is true and then given that the alternative $\mu = 295$ is true.

(a) Given that $\mu = 299$, then $\bar{x} \sim N(299, 3/\sqrt{6})$. So we must now compute $P(\bar{x} < 297.985)$. To do so we can use the **NORMDIST** program or the **normalcdf(** command where $P(\bar{x} < 297.985) \approx P(-1E99 \leq \bar{x} \leq 297.985)$. Either way, we see that the power against the alternative $\mu = 299$ is .2036.

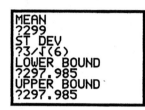 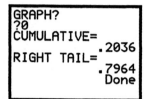 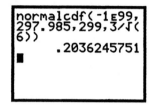

(b) Given that $\mu = 295$, then $\bar{x} \sim N(295, 3/\sqrt{6})$. We simply recalculate the probability with this new value of the mean.

We see that the power against the alternative $\mu = 295$ is .9926. Obviously if $\mu = 295$, then we would be very likely to reject H_0: $\mu = 300$ in favor of the alternative H_a: $\mu < 300$.

Exercise 6.70. (a) An SRS of size 72 is taken from a population having $\sigma = 15$ to test the hypothesis H_0: $\mu = 128$ versus a two-sided alternative at the 5% level of significance. Find the power against the alternative $\mu = 134$.

Solution: Again we must first find the rejection regions. For a two-sided alternative at the 5% level of significance, we allow 2.5% at each tail. Thus we reject if the z-test statistic is beyond ± 1.96. That is, we reject if

$$\frac{\bar{x} - 128}{15/\sqrt{72}} < -1.96 \quad \text{or} \quad \frac{\bar{x} - 128}{15/\sqrt{72}} > 1.96.$$

Equivalently, we reject if $\bar{x} < 124.535$ or if $\bar{x} > 131.465$. Now assuming that $\mu = 134$, then $\bar{x} \sim N(134, 15/\sqrt{72})$. We must now compute $P(\bar{x} < 124.535) + P(\bar{x} > 131.465)$, which is equivalent to $1 - P(124.535 \leq \bar{x} \leq 131.465)$. Using either of our methods, we see that the power against the alternative $\mu = 134$ is .9242.

```
1-normalcdf(124.
535,131.465,134;
15/√(72))
        .924215658
```

```
MEAN
?134
ST DEV
?15/√(72)
LOWER BOUND
?124.535
UPPER BOUND
?131.465
```

```
GRAPH?
?0
PROB=
            .0758
            Done
1-.0758
            .9242
■
```

Exercise 6.71. Using the result of Exercise 6.67, give the probability of a Type I error and the probability of a Type II error for the alternative hypothesis of H_a: $\mu = 295$.

Solution: The significance level $\alpha = .05$ is the probability of a Type I error. Since the power against the alternative $\mu = 295$ is .9926, the probability of a Type II error for this alternative is $1 - .9926 = .0074$.

CHAPTER
7

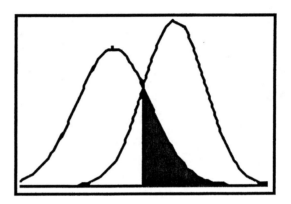

Inference for
Distributions

In this chapter we introduce the t distributions and the various t procedures that are used for confidence intervals and significance tests about the mean of a normal population for which the standard deviation is unknown.

7.1 Inference for the Mean of a Population

We begin with the short program **TSCORE** that allows us to find critical values (*t* scores) upon specifying the degrees of freedom and confidence level.

```
PROGRAM:TSCORE              :"tcdf(0,X,M)"→Y₁
:Disp "DEG. OF FREEDOM"     :solve(Y₁-R/2,X,2)→Q
:Input M                    :Disp "T SCORE"
:Disp "CONF. LEVEL"         :Disp round(Q,3)
:Input R
```

Exercise 7.4. Find the critical values t^* for confidence intervals for the mean in the following cases: (a) a 95% confidence interval with $n = 10$ observations, (b) a 99% confidence interval from an SRS of 20 observations, (c) an 80% confidence interval from a sample of size 7.

Solution: We recall that the confidence intervals are based on t distributions with $n-1$ degrees of freedom. For part (a), we need 9 degrees of freedom. Below are the outputs from running the **TSCORE** program for each part.

```
DEG. OF FREEDOM        DEG. OF FREEDOM        DEG. OF FREEDOM
?9                     ?19                    ?6
CONF. LEVEL            CONF. LEVEL            CONF. LEVEL
?.95                   ?.99                   ?.8
T SCORE                T SCORE                T SCORE
          2.262                  2.861                  1.44
          Done                   Done                   Done
```

We now examine confidence intervals for one mean that will require the **TInterval** feature from the **STAT TESTS** menu.

7.16 Sharks. Here are the lengths in feet for 44 great white sharks.

18.7	12.3	18.6	16.4	15.7	18.3	14.6	15.8	14.9	17.6	12.1
16.4	16.7	17.8	16.2	12.6	17.8	13.8	12.2	15.2	14.7	12.4
13.2	15.8	14.3	16.6	9.4	18.2	13.2	13.6	15.3	16.1	13.5
19.1	16.2	22.8	16.8	13.6	13.2	15.7	19.7	18.7	13.2	16.8

Give a 95% confidence interval for mean length of great white sharks. Based on this interval, is there significant evidence at the 5% level to reject the claim "Great white sharks average 20 feet in length?"

Solution: First enter the data into a list, say list **L1**. Next press **STAT**, scroll right to **TESTS**, then scroll down and enter **TInterval** (or press **8**). Set the **Inpt** to **Data**, the list to **L1**, the **C-Level** to .95, and press **ENTER** on **Calculate**. The sample mean, sample deviation, and confidence interval are all displayed.

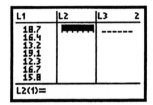 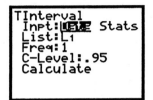 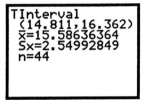

Since the value of 20 is not in the 95% confidence interval (14.811, 16.362), we do have significant evidence at the 5% level to reject the claim that great white sharks average 20 feet in length.

7.17 Calcium and blood pressure. In an experiment on the effect of calcium on blood pressure, a control group of 27 members received a placebo. For this group, the mean systolic blood pressure was $\bar{x}$ = 114.9 with a sample deviation of s = 9.3. Give a 95% confidence interval for the mean blood pressure in the population from which this group was recruited.

Solution: Call up the **TInterval** feature from the **STAT TESTS** menu. Set the **Inpt** by pressing **ENTER** on **Stats**. Then enter the values of $\bar{x}$, **Sx**, **n**, and **C-Level**, and press **ENTER** on **Calculate**. We obtain a confidence interval of (111.22, 118.58).

We now perform some significance tests about the mean using the **T-Test** feature (item **2**) from the **STAT TESTS** menu.

7.12 Will they charge more? A bank has omitted the credit card fee on an SRS of 200 customers to see if there will be an increase in the amounts the customers charge over the course of a year. The mean increase was $332 with a standard deviation of $108. Is there significant evidence at the 1% level that the mean amount charged has increased? Give a 99% confidence interval for the mean amount of increase.

Solution: We shall test H_0: $\mu = 0$ with the alternative H_a: $\mu > 0$. Press **STAT**, scroll right to **TESTS**, then press **2** for **T-Test**. Set the **Inpt** to **Stats**, adjust the settings, and calculate.

We obtain a P-value of 0. If there were no change in average, then there would be no chance of obtaining a sample mean increase as high as $332. We therefore have significant evidence to reject H_0, even at the 1% level, in favor of the alternative.

Next, execute the **TInterval** feature (the statistics should already be set from the previous calculation). The true mean increase should be from $312.14 to $351.86.

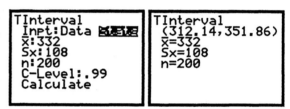

7.13   Auto crankshafts. Here are the measurements (in millimeters) of a critical dimension on a sample of 16 auto engine crankshafts.

224.120	224.001	224.017	223.982	223.989	223.961
223.960	224.089	223.987	223.976	223.902	223.980
224.098	224.057	223.913	223.999		

The dimension is supposed to be 224 mm and the variability of the manufacturing process is unknown. Is there evidence that the mean dimension is not 224 mm? State the appropriate H_0 and H_a and carry out a t test. Give the P-value and conclusion.

Solution: We shall test the null hypothesis H_0: $\mu = 224$ with the two-sided alternative H_a: $\mu \neq 224$. First enter the data into a list, say list **L2**. Then press **STAT**, scroll right to **TESTS**, and press **2** for **T-Test**. Set the **Inpt** to **Data**, adjust the settings, and calculate.

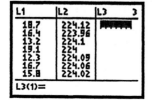 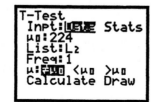 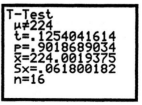

We obtain a *P*-value of .902. If H_0 were true, there would be 90% chance of obtaining a sample mean as far away from 224 as $\bar{x}$ = 224.0019. This high *P*-value means that we do not have evidence that the mean dimension is not 224 mm.

We next work an exercise that uses the matched pair *t* procedure.

7.9 Right versus left. We wish to assess whether right-handed subjects can complete a task with a right-hand thread significantly faster than a task with a left-hand thread. The times in seconds follow.

Subject	Right	Left	Subject	Right	Left
1	113	137	14	107	87
2	105	105	15	118	166
3	130	133	16	103	146
4	101	108	17	111	123
5	138	115	18	104	135
6	118	170	19	111	112
7	87	103	20	89	93
8	116	145	21	78	76
9	75	78	22	100	116
10	96	107	23	89	78
11	122	84	24	85	101
12	103	148	25	88	123
13	116	147			

State H_0 and H_a and carry out a matched pair *t* test. Give the *P*-value and conclusion.

Solution: By considering the average difference μ in times, we will test the null hypothesis H_0: $\mu = 0$, which asserts that there is no difference in times. We must choose a one-sided alternative that is equivalent to the right-hand thread times being faster. In other words, the alternative is that the average left-hand thread time is higher, which means that $\mu < 0$. The alternative is then H_a: $\mu < 0$.

We must now enter the *differences* of the measurements into a list, say list **L3**. We can enter the values into the list as $113-137$, $105-105$, etc., and the calculator will do the subtraction for us when we press **ENTER**. After entering the data, execute the **T-Test** command from the **STAT TESTS** menu.

L1	L2	L3	3
18.7	224.12	-24	
16.4	223.96	0	
13.2	224.1	-3	
19.1	224	-7	
12.3	224.09	23	
16.7	224.06	-52	
15.8	224.02	-16	

L3 =(-24, 0, -3, -7...

```
T-Test
 Inpt:Data Stats
 μ0:0
 List:L3
 Freq:1
 μ:≠μ0  <μ0  >μ0
 Calculate  Draw
```

```
T-Test
 μ<0
 t=-2.903732389
 P=.0038957552
 x̄=-13.32
 Sx=22.93599791
 n=25
 ■
```

We obtain a *P*-value of .0039; thus we can say that the left-hand times average significantly higher. If the averages were the same, then there would be only a .0039 chance of having an average difference in times as low as $\bar{x} = -13.32$ seconds.

7.26 Optional: The power of a *t* test.   The bank in Exercise 7.12 wants to be quite certain of detecting a mean increase of $\mu = \$100$ in the amount charged at the $\alpha = .01$ significance level. Find the approximate power of the one-sided test H_0: $\mu = 0$ with a sample of only $n = 50$ customers against the alternative $\mu = \$100$. Use the value of $\sigma = 108$ (an estimate based on the data from Exercise 7.12).

Solution: We first find the rejection region of the test at the .01 level of significance. For a one-sided alternative, we allow .01 probability at the right tail, which corresponds to the critical value t^* of a 98% confidence interval using the $t(n-1) = t(49)$ distribution.

Using the **TSCORE** program, we find that the t^* value is 2.405.

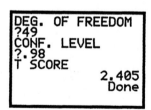

```
DEG. OF FREEDOM
?49
CONF. LEVEL
?.98
T SCORE
          2.405
          Done
```

Thus, we reject H_0 if

$$\frac{\bar{x} - 0}{108 / \sqrt{50}} > 2.405.$$

Equivalently, we reject if $\bar{x} > 36.7328$. We must then find the probability that $\bar{x}$ is greater than 36.7328 given that the alternative $\mu = 100$ is true.

Given that $\mu = 100$ and $\sigma = 108$, then $\bar{x} \sim N(100,\ 108 / \sqrt{50})$. Using this distribution, we must compute $P(\bar{x} > 36.7328)$.

For this calculation, we use the built-in **normalcdf(** command from the **DISTR** menu. We see that the power against the alternative $\mu = 100$ is essentially 1. If $\mu = 100$, then we are certain to reject the hypothesis H_0: $\mu = 0$ with a sample of size 50.

7.2 Comparing Two Means

We next consider confidence intervals and significance tests for the difference between means of two independent normal populations that both have unknown standard deviations. The results are based on independent random samples, of sizes n_1 and n_2 respectively, taken from each population. We will now use the **2-SampTInt** and the **2-SampTTest** features from the **STATS TESTS** menu.

These features require that we specify whether or not we wish to use a "pooled variance" s_p^2. We should specify "Yes" only under the assumption that the two populations have the *same* (unknown) variance. In this case, the critical values t^* are obtained from the $t(n_1 + n_2 - 2)$ distribution.

When we do not specify a pooled variance, then the standardized distribution given by

$$\frac{(\bar{x}_1 - \bar{x}_2) - (\mu_1 - \mu_2)}{\sqrt{\dfrac{s_x^2}{n_1} + \dfrac{s_y^2}{n_2}}}$$

is only an approximate t distribution. In this case, the degrees of freedom r are given by

$$r = \frac{\left(\dfrac{s_x^2}{n_1} + \dfrac{s_y^2}{n_2}\right)^2}{\dfrac{1}{n_1 - 1}\left(\dfrac{s_x^2}{n_1}\right)^2 + \dfrac{1}{n_2 - 1}\left(\dfrac{s_y^2}{n_2}\right)^2}.$$

When r is not an integer, the critical values t^* are obtained from the $t(n)$ distribution, where n is the greatest integer less than r.

7.41 Talented 13-year-olds. Here are the Math SAT scores from tests given to 13-year-olds by the Johns Hopkins Regional Talent Searches. Give a 99% confidence interval for the difference between the mean score for males and the mean score for females in the population that Johns Hopkins searches.

Group	n	$\bar{x}$	s
Males	19,833	416	87
Females	19,937	386	74

Solution: Call up the **2-SampTInt** feature (item **0**) from the **STAT TESTS** menu and set the **Inpt** to **Stats**. Then enter the given statistics and enter **No** for **Pooled**. Then press **ENTER** on **Calculate**.

```
2-SampTInt          2-SampTInt          2-SampTInt
 Inpt:Data STATS    ↑n1:19833            (27.913,32.087)
 x̄1:416             x2:386              df=38706.44937
 Sx1:87             Sx2:74              x̄1=416
 n1:19833           n2:19937            x̄2=386
 x̄2:386             C-Level:.99          Sx1=87
 Sx2:74             Pooled:NO Yes       ↓Sx2=74
↓n2:19937■          Calculate           ■
```

We see that the difference in means is from 27.913 to 32.087. For this population of 13-year-olds, males average from 27.913 points higher to 32.087 points higher than females on the Math SAT.

7.44 IQ scores for boys and girls. Here are the IQ test scores of 31 seventh-grade girls in a midwest school district.

114	100	104	89	102	91	114	114	103	105	
108	130	120	132	111	128	118	119	86	72	
111	103	74	112	107	103	98	96	112	112	93

The IQ test scores of 47 seventh-grade boys in the same district follow.

111	107	100	107	115	111	97	112	104	106	113
109	113	128	128	118	113	124	127	136	106	123
124	126	116	127	119	97	102	110	120	103	115
93	123	79	119	110	110	107	105	105	110	77
90	114	106								

Give a 90% confidence interval for the difference between the mean IQ scores of all boys and girls in the school district.

Solution: Let μ_1 be the mean score of all boys in the district, and let μ_2 be the mean score of all girls. We shall find a confidence interval for $\mu_1 - \mu_2$.

We must first enter the data into lists in the **STAT EDIT** screen. Clear any existing data, and enter the girls' scores into **L1** and the boys' scores into **L2**. Call up the **2-SampTInt** feature from the **STAT TESTS** menu and set the **Inpt** to **Data**.

Since the boys' scores are in **L2**, set **List1** to **L2** and set **List2** to **L1**. Our result then will be for mean boy score minus mean girl score. Enter **No** for **Pooled**, then press **ENTER** on **Calculate**.

```
L1      L2      L3     3
114     111
108     109
111     124
100     93
130     90
103     107
104     113
L3(1)=
```

```
2-SampTInt
Inpt:DATA Stats
List1:L2
List2:L1
Freq1:1
Freq2:1
C-Level:.9
↓Pooled:NO Yes
```

```
2-SampTInt
(-.0878,10.325)
df=56.93171263
x̄1=110.9574468
x̄2=105.8387097
Sx1=12.1206928
↓Sx2=14.2714091
```

We find that $-.0878 \le \mu_1 - \mu_2 \le 10.325$. That is, the boys could average from .0878 points lower to 10.325 points higher than the girls. In particular, the averages could be the same since 0 is in the confidence interval.

7.47 Will they charge more? A bank has offered different credit card plans to two sample groups to see if there is a significant difference in the average amounts the groups charge per year. The summary statistics follow. Test for a difference in averages.

Group	n	$\bar{x}$	s
A	150	$1987	$392
B	150	$2056	$413

Solution: Let μ_1 be the mean amount charged by all those offered Plan A, and let μ_2 be the mean amount charged by all those in Plan B. We shall test H_0: $\mu_1 = \mu_2$ versus the alternative H_a: $\mu_1 \ne \mu_2$.

Call up the **2-SampTTest** feature (item 4 in the **STAT TESTS** menu) and set the **Inpt** to **Stats**. Enter the given statistics, set the alternative, enter **No** for **Pooled**, and calculate.

```
2-SampTTest          2-SampTTest          2-SampTTest
 Inpt:Data STATS     ↑n1:150              μ1≠μ2
 x̄1:1987             x̄2:2056             t=-1.484110074
 Sx1:392             Sx2:413              p=.1388395127
 n1:150              n2:150               df=297.1921082
 x̄2:2056            μ1:≠μ2 <μ2 >μ2       x̄1=1987
 Sx2:413             Pooled:No Yes        ↓x̄2=2056
↓n2:150              Calculate Draw       ■
```

With a P-value of .1388, we conclude that there is no real statistical difference in the average amounts charged. If the difference in averages were 0, there still would be a 13.88% chance of the sample means being as far apart as they are with samples of these sizes.

If we calculate a 95% confidence interval with the **2-SampTInt** feature, we see that $-160.50 \le \mu_1 - \mu_2 \le 22.49$. That is, the mean from Group A could be from $160.50 lower to $22.49 higher than the mean from Group B. Since $0 is in this confidence interval, the means could be equal, which again verifies no statistical difference.

```
2-SampTInt
 (-160.5,22.496)
 df=297.1921082
 x̄1=1987
 x̄2=2056
 Sx1=392
↓Sx2=413
■
```

7.43 Active versus passive learning. A study of two styles of computer-assisted learning is to be tested to see if "active" learning is superior to "passive" learning. Here are the scores on a quiz by 24 children in the Active group.

29	28	24	31	15	24	27	23	20	22	23	21
24	35	21	24	44	28	17	21	21	20	28	16

Here are the scores by the children in the Passive group.

16	14	17	15	26	17	12	25	21	20	18	21
20	16	18	15	26	15	13	17	21	19	15	12

Is there good evidence that active learning is superior to passive learning? State hypotheses, give a P-value, and state a conclusion.

Solution: Let μ_1 be the mean quiz score of all those in active learning, and let μ_2 be the mean quiz score of all those in passive learning. We shall test H_0: $\mu_1 = \mu_2$ versus the alternative H_a: $\mu_1 > \mu_2$.

Now enter the data into lists, say **L3** for the Active group and **L4** for the Passive group. Then call up the **2-SampTTest** feature and set the **Inpt** to **Data**. Since the Active scores are in **L3**, set **List1** to **L3**. Then our result will be for mean Active score minus mean Passive score. Enter **No** for **Pooled**, then press **ENTER** on **Calculate**.

We obtain an extremely low P-value of .0000583. Thus there is significant evidence to reject H_0. If the true means were equal, then there would be almost no chance of $\bar{x}_1$ being so much higher than $\bar{x}_2$. We therefore reject H_0 and conclude that the mean score from Active learning is higher than the mean score from Passive learning.

7.3 Inference for Population Spread

We now discuss the F-ratio test, a test for determining whether or not two normal populations have the same variance. If so, then we would be justified in using the pooled two-sample t procedures for confidence intervals and significance tests about the difference in means.

7.57 Studying speech. The data for VOTs of children and adults follow. Test whether or not there is a significant difference in the standard deviations.

Group	n	$\bar{x}$	s
Children	10	−3.67	33.89
Adults	20	−23.17	50.74

Solution: Let σ_1 be the standard deviation of all children in the population under study, and let σ_2 be the standard deviation of all adults. We shall test H_0: $\sigma_1 = \sigma_2$ versus H_a: $\sigma_1 \neq \sigma_2$. The null hypothesis is equivalent to testing whether or not the ratio σ_2/σ_1 equals 1. We note that the sample deviation ratio of adults to children is $50.74/33.89 \approx 1.497$ and the ratio of children to adults is $33.89/50.74 \approx .668$.

Call up the **2-SampFTest** feature (item **D**) from the **STAT TESTS** menu and set the **Inpt** to **Stats**. Enter the statistics and alternative, then calculate.

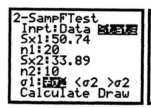

With a high P-value of .2152, we can say that there is no clear significant difference in standard deviations. If σ_1 were equal to σ_2, then there still would be a 21.52% chance of Sx1/Sx2 being as far away from 1 as 1.497.

7.58 Students' attitudes. The SSHA scores from Exercise 7.36 follow.

Women's scores

154	109	137	115	152	140	154	178	101
103	126	126	137	165	165	129	200	148

Men's scores

108	140	114	91	180	115	126	92	169	146
109	132	75	88	113	151	70	115	187	104

Test whether or not the spread of SSHA scores is different among men and women.

Solution: First enter the data into lists, say **L1** for women and **L2** for men. Then call up the **2-SampFTest** feature from the **STAT TESTS** menu and set the **Inpt** to **Data**. Enter the appropriate lists and alternative, then calculate.

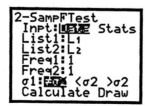

We note that the sample deviation ratio of men to women is $Sx2/Sx1 \approx$ $32.852/26.436 \approx 1.24$. With a P-value of .3725, we can say that there is no significant difference in standard deviations. If σ_1 were equal to σ_2, there would be greater than a 37% chance of the sample deviation ratio being as far away from 1 as it is.

CHAPTER
8

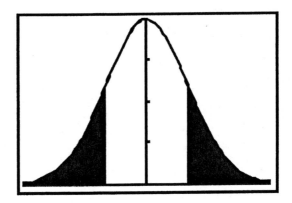

Inference for Proportions

8.1 Inference for a Population Proportion
8.2 Comparing Two Proportions

We now discuss how to use the TI-83 to find confidence intervals and to conduct hypothesis tests for a single proportion and for the difference in two proportions.

8.1 Inference for a Population Proportion

We begin by considering an unknown population proportion p. To estimate p, we conduct a large SRS of size n and find the sample proportion $\hat{p}$. We then create a confidence interval by adding and subtracting the margin of error $z^*\sqrt{\hat{p}(1-\hat{p})/n}$, where z^* is the appropriate critical value depending on the level of confidence.

Confidence Interval for a Proportion

To compute a confidence interval for a proportion, we can use the TI-83's built-in **1-PropZInt** feature (item **A**) in the **STAT TESTS** menu.

8.18 Unhappy HMO patients. Out of 639 complaints filed with a health maintenance organization, 54 of the complainers left the HMO voluntarily. If this group is considered a random sample of all future complainers, find a 90% confidence interval for the true proportion of complainers who voluntarily leave the HMO.

Solution: Press **STAT**, scroll right to **TESTS**, then scroll down to **1-PropZInt** (or press **APLHA A**) and press **ENTER**.

Enter the data, 54 for **x**, 639 for **n**, and the desired level of confidence. Then scroll down to **Calculate** and press **ENTER**.

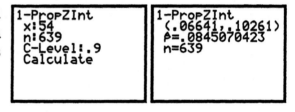

The 90% confidence interval (.06641, .10261) is given along with the sample proportion $\hat{p} = .0845$.

The PSAMPSZE Program

As with confidence intervals for the mean, we would often like to know in advance what sample size would provide a certain maximum margin of error m with a certain level of confidence. The required sample size is given by

$$n \geq \left(\frac{z^*}{m}\right)^2 p^*(1-p^*),$$

where z^* is the appropriate critical value depending on the level of confidence and p^* is a guessed value of the true proportion p. If $p^* = .50$, then the resulting sample size ensures that the margin of error is no more than m regardless of the true value of p.

The program **PSAMPSZE** computes the required sample size after one enters the desired error m, the confidence level, and the guess p^*.

```
PROGRAM:PSAMPSZE
:Disp "DESIRED ERROR"           :If int(M)=M
:Input E                        :Then
:Disp "CONF. LEVEL"             :M→N
:Input R                        :Else
:Disp "GUESS OF P"              :int(M+1)→N
:Input P                        :End
:invNorm((R+1)/2,0,1)→Q         :Disp "SAMPLE SIZE="
:(Q/E)²*P(1-P)→M                :Disp int(N)
```

8.19 Do you go to church?   In a Gallup Poll of 1785 adults, 750 responded "Yes" to the question "Did you attend church or synagogue in the last 7 days?" Give a 99% confidence interval for the proportion of all U.S. adults who attended church or synagogue during the week preceding the poll. How large a sample would be required to obtain a maximum margin of error of .01 with 99% confidence?

Solution: First, to find the confidence interval, press **STAT**, scroll right to **TESTS**, then scroll down to **1-PropZInt** and press **ENTER**.

Enter the data and the desired C-Level, then scroll down to **Calculate** and press **ENTER**.

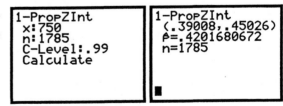

We obtain an interval of (0.39, 0.45) with $\hat{p} = .42$, which can serve as our guess p^* of the true proportion p. This interval provides strong evidence that less than half the population attended church or synagogue in that week.

Now, to find the required sample size, we execute the **PSAMPSZE** program. We find that an SRS of size 16,163 would be required to have a maximum margin of error of only .01 with 99% confidence.

```
PrgmPSAMPSZE
DESIRED ERROR
?.01
CONF. LEVEL
?.99
GUESS OF P
?.42■
```

```
CONF. LEVEL
?.99
GUESS OF P
?.42
SAMPLE SIZE=
            16163
            Done
```

Significance Tests

We now conduct hypothesis tests for a single population proportion p using the **1-PropZTest** feature (item 5) in the **STAT TESTS** menu.

8.11 (b) We want to be rich. In a recent year, 73% of first-year college students identified "being very well-off financially" as an important personal goal. In a survey of its first-year students, a state university finds that 132 of 200 respondents say that this goal is important. Is there good evidence that the proportion of first-year students at this university who think being very well off is important differs from the national value? State hypotheses and give a P-value and conclusion.

Solution: The sample proportion for this university's first-year students is $\hat{p}$ = 132/200 = .66. Since this value is much lower than the national proportion, we might think that this university's true proportion p is lower than .73. Thus we shall test the claim H_0: $p = .73$, with the alternative H_a: $p < .73$.

Press **STAT**, scroll right to **TESTS**, then press **5** for the **1-PropZTest** feature. Enter .73 for p_0, then enter 132 for **x** and 200 for **n**. Enter the desired alternative, then press **ENTER** on **Calculate**.

The z statistic is given as −2.23 and the P-value is .01288. If p were equal to .73, then there would be only a 1.288% chance of $\hat{p}$ being as small as .66. We have strong evidence to reject H_0 in favor of H_a.

```
1-PropZTest
p0:.73
x:132
n:200
prop≠p0  <p0  >p0
Calculate Draw
```

```
1-PropZTest
prop<.73
z=-2.229819585
p=.0128796682
p̂=.66
n=200
```

8.15 Stolen Harleys. Harley-Davidson motorcycles make up 14% of all motorcycles registered in the United States. In 1995, 9224 motorcycles were reported stolen and of these 2490 were Harleys. Consider the results of 1995 to be sample of recent years. Is the proportion of Harleys among stolen bikes significantly higher than their share of all motorcycles?

Solution: Let p be the true proportion of Harleys stolen. We shall test the claim H_0: $p = .14$ versus the alternative H_a: $p > .14$.

We use the **1-PropZTest** feature to obtain a P-value of 0. If p were equal to .14, then there would be no chance of $\hat{p}$ being as high as .27. There is strong evidence to reject H_0.

```
1-PropZTest
P₀:.14
x:2490
n:9224
prop≠P₀ <P₀ ▓F₀
Calculate Draw
```

```
1-PropZTest
prop>.14
z=35.96796574
P=0
p̂=.2699479618
n=9224
```

Using the **1-PropZInt** feature, we can quickly find a 99% confidence interval for p, which gives further evidence that $p > .14$.

```
1-PropZInt
x:2490
n:9224
C-Level:.99
Calculate
```

```
1-PropZInt
(.25804,.28185)
p̂=.2699479618
n=9224
```

8.2 Comparing Two Proportions

We now study confidence intervals and significance tests for the difference in two subpopulation proportions p_1 and p_2. We base the results on independent SRSs, of sizes n_1 and n_2 respectively, that yield sample proportions $\hat{p}_1$ and $\hat{p}_2$. We shall use the TI-83's **2-PropZInt** and **2-PropZTest** features in the **STAT TESTS** menu for the necessary calculations.

8.26 Free speech. In a 1958 study on the influence of religion, one question asked if the right of free speech included the right to make speeches in favor of communism. Of 267 Protestants responding, 104 said "Yes," while 75 out of 230 Catholics said "Yes." Give a 95% confidence interval for the difference in the true proportions among all Protestants and Catholics who agreed that communist speeches are protected.

Solution: Let p_1 be the true proportion of all Protestants who agreed, and let p_2 be the true proportion of all Catholics who agreed. We shall find a confidence interval for $p_1 - p_2$.

Press **STAT**, scroll right to **TESTS**, then scroll down to the **2-PropZInt** feature (or press **ALPHA B**). Enter the statistics, then press **ENTER** on **Calculate**.

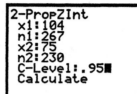

The confidence interval $(-.0208, .14764)$ is displayed along with the sample proportions, which are respectively $\hat{p}_1 = .3895$ and $\hat{p}_2 = .326$. Thus with 95% confidence, we can say that $-0.0208 \leq p_1 - p_2 \leq 0.14764$.

That is, the true proportion among Protestants could be from 2.08 percentage points lower to 14.76 percentage points higher than the true proportion among Catholics.

8.33 Are urban students more successful? In a study of chemical engineering students at North Carolina State University, the backgrounds of students successful in a course were checked. Of 65 students with urban or suburban backgrounds, 52 succeeded. Of 55 students with rural or small-town backgrounds, 30 succeeded. (a) Is there good evidence that the proportions of students who succeed are different among these two types of students? (b) Give a 90% confidence interval for the difference in the true proportions who succeed in the course among students with these two types of backgrounds.

Solution: (a) Let p_1 be the true proportion of all urban/suburban students who succeed in the course, and let p_2 be the true proportion of all rural/small-town students who succeed. Then $\hat{p}_1 = 52/65 = .80$, and $\hat{p}_2 = 30/55 \approx .545$. Since $\hat{p}_1$ is so much higher than $\hat{p}_2$, we shall test the claim H_0: $p_1 = p_2$ versus the alternative H_a: $p_1 > p_2$. To do so, we use the **2-PropZTest** feature (item 6) from the **STAT TESTS** menu.

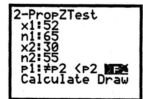

We obtain a P-value of .0014. That is, if the true proportions among these two types of students were equal, then there would be only a .0014 probability of $\hat{p}_1$ being so much larger than $\hat{p}_2$. This low P-value gives strong evidence to reject the claim $p_1 = p_2$ in favor of the alternative $p_1 > p_2$.

(b) To find a confidence interval, we now use the **2-PropZInt** feature.

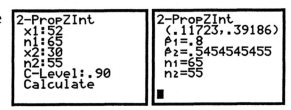

The confidence interval is given as (.11723, .39186). That is, the true proportion among urban/suburban students who succeed in the course is from 11.72 percentage points higher to 39.19 percentage points higher than the true proportion among rural/small-town students who succeed.

8.34 Small-business failures. In a study of small-business failures in central Indiana, 15 out of 106 businesses headed by men failed, while 7 out of 42 businesses headed by women failed. Is there a significant difference in the rates at which businesses headed by men and women fail?

Solution: Let p_1 be the true failure rate for businesses headed by men and let p_2 be the true failure rate for businesses headed by women. We shall test the claim H_0: $p_1 = p_2$ versus the alternative H_a: $p_1 \neq p_2$.

Adjust the settings in the **2 - PropZTest** feature and calculate.

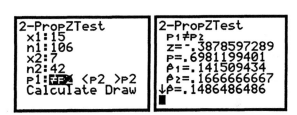

With a large P-value of .698, we can say that there is no significant difference in the failure rates. If they were the same, then there would still be a 69.8% chance of $\hat{p}_1$ and $\hat{p}_2$ being as far apart as they are with samples of these sizes.

CHAPTER
9

Inference for
Two-Way Tables

In this chapter we describe how to use the TI-83 to perform a chi-square test on data in a two-way table. We test whether the categories of traits listed as the row variables are related to the traits listed as the column variables or whether row and column traits are independent.

9.1 Two-Way Tables

We previously examined two-way tables in Section 2.5 using the **TWOWAY** program. Let us quickly review that program.

9.1 Extracurricular activities and grades. North Carolina State University studied student success in a certain chemical engineering course. One factor studied was the number of hours per week a student spent on extracurricular activities. Here are the data.

Grade	Extracurricular activities (hours per week)		
	< 2	2 to 12	> 12
C or better	11	68	3
D or F	9	23	5

Find the proportion of successful students (C or better) in each of the three extracurricular activity groups.

Solution: Press **MATRX**, scroll right to **EDIT**, and press **1**. Enter the dimensions for matrix **[A]** as 2 × 3. Then enter the data into the matrix.

```
MATRIX[A]  2 ×3
[ 11   68   3  ]
[ 9    23   5  ]
```

Execute the **TWOWAY** program. Then call up matrix **[B]** on the Home screen (scroll right to see the remainder of the matrix) or view matrix **[B]** from the **MATRX EDIT** screen. Recall that matrix **[B]** gives the proportions among the whole group that was sampled.

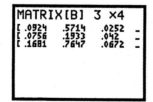

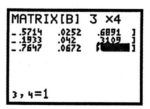

The last column and last row give the totals. Among the whole group sampled, 68.91% made C or better, and 31.09% made a D or F. Of this group of students, 16.81% spend less than 2 hours per week on extracurricular activities, while 76.47% spend from 2 to 12 hours per week.

Now view matrix [C], which gives the conditional proportions for each of the three extracurricular activity groups.

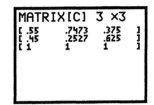

```
MATRIX[C] 3 ×3
[ .55    .7473   .375  ]
[ .45    .2527   .625  ]
[ 1      1       1     ]
```

Among students who spend less than 2 hours per week, 55% succeeded in the course. Among students who spend 2 to 12 hours per week, 74.73% succeeded in the course. Among students who spend more than 12 hours per week, only 37.5% succeeded in the course. There seems to be a significant difference in these proportions.

(Note that matrix [D] gives the proportions of the extracurricular activity groups for each type of grade category.)

Observed versus Expected Counts

The TI-83 has a built-in feature, the X^2-Test (item C) in the STAT TESTS menu, that will compute the expected counts under the assumption that all the proportions for each category type are the same.

To access this feature, press STAT, scroll right to TESTS, then scroll down to X^2-Test and press ENTER (or press ALPHA C). Enter matrix [A] for Observed (to do so, press MATRX, then press 1), then enter matrix [E] for Expected (to do so, press MATRX, then press 5). Press ENTER on Calculate, then compare matrix [E] with matrix [A].

```
X²-Test
 Observed: [A]
 Expected: [E]
 Calculate Draw
```
```
MATRIX[E] 2 ×3
[ 13.782  62.706  5.5126 ]
[ 6.2185  28.294  2.4874 ]
```
```
MATRIX[A] 2 ×3
[ 11   68   3  ]
[ 9    23   5  ]
```

If the proportions of students who succeed were the same for each extracurricular activity group, then we would expect to average respectively 13.782, 62.706, and 5.5126 students who succeed in each of these categories. These expected counts differ slightly from the observed counts of 11, 68, and 3.

9.2 Smoking by students and their parents.
How are the smoking habits of students related to their parents' smoking? Here are data from a survey of students in eight Arizona high schools.

	Student smokes	Student does not smoke
Both parents smoke	400	1380
One parent smokes	416	1823
Neither parent smokes	188	1168

Calculate the proportion of students who smoke in each of the three parent groups.

Solution: We first enter the data into a 3 × 2 matrix **[A]** and execute the **TWOWAY** program. Then view matrices **[B]** and **[D]**.

```
MATRIX[B]  4 ×3
[ .0744   .2567   .3312   ]
[ .0774   .3392   .4166   ]
[ .035    .2173   .2523   ]
[ .1868   .8132   1       ]
```

```
MATRIX[D]  3 ×3
[ .2247   .7753   1   ]
[ .1858   .8142   1   ]
[ .1386   .8614   1   ]
```

Analyzing matrix **[B]** (the overall percents): Among the whole group sampled, both parents of 33.12% smoke, while neither parent of 25.23% smokes. Of this group of students, 18.68% smoke and 81.32% do not.

Analyzing matrix **[D]** (the column percents): Among students whose parents both smoke, 22.47% also smoke. Among students with one parent who smokes, 18.58% smoke. But among students neither of whose parents smokes, only 13.86% smoke. There seems to be a clear difference in these proportions.

(Note that matrix **[C]** gives the proportions of the parent groups for each type of smoking/nonsmoking student.)

Now execute the X^2-Test feature and compare the observed counts versus the expected counts.

```
MATRIX[E]  3 ×2
[ 332.49   1447.5   ]
[ 418.22   1820.8   ]
[ 253.29   1102.7   ]
```

```
MATRIX[A]  3 ×2
[ 400   1380   ]
[ 416   1823   ]
[ 188   1168   ]
```

If the proportions of students who smoke were the same for each parent group, then we would expect to average respectively 332.49, 418.22, and 253.29 students who smoke in each of these categories. Two of these expected counts differ greatly from the observed counts of 400, 416, and 188.

9.2 The Chi-Square Test

We now see how to test whether all of the proportions of a row (column) trait are the same regardless of the column (row) trait.

The X^2-Test Feature

To test whether all the proportions are the same, we again use the X^2-Test feature (item **C**) from the **STAT TESTS** menu.

9.4 Smoking by students and their parents.   Use the two-way table from Exercise 9.2 to test whether the proportions of students who smoke are the same regardless of parent group.

Solution: The null hypothesis H_0 is that proportions of students who smoke are the same regardless of parent group.

Looking at matrix **[D]** from the **TWOWAY** program, we see that the percentages who smoke are respectively 22.47%, 18.58%, and 13.86%. These percentages are quite different, so perhaps we have evidence to reject H_0 in favor of the alternative that states that the proportions of students who smoke are not the same for each parent group. In other words, at least one parent group type would differ from the others. The proportions of all three parent group types might differ, or two might be the same while the third differs, but at least one differs.

We will now test H_0 with the chi-square test. Press **STAT**, scroll right to **TESTS**, then press **ALPHA C** (or scroll down) for the X^2-Test feature. Enter matrix **[A]** for the **Observed** and enter matrix **[E]** for the **Expected**. Then scroll down to **Calculate** and press **ENTER**.

```
EDIT CALC TESTS
9↑2-SampZInt…
0:2-SampTInt…
A:1-PropZInt…
B:2-PropZInt…
■X²-Test…
D:2-SampFTest…
E↓LinRegTTest…
```

```
X²-Test
Observed: [A]
Expected: [E]
Calculate Draw
```

```
X²-Test
X²=37.56634178
P=6.9594117E-9
df=2
```

We obtain a chi-square test statistic of 37.566 and a P-value of .000000006959 which is essentially 0. Now let us again view matrix [A] and compare it to matrix [E].

```
MATRIX[A] 3 ×2          MATRIX[E] 3 ×2
[ 400    1380  ]        [ 332.49   1447.5  ]
[ 416    1823  ]        [ 418.22   1820.8  ]
[ 188    1168  ]        [ 253.29   1102.7  ]
```

We can interpret the P-value as follows: If the proportions of students who smoke were the same regardless of the parent group, then there would be almost no chance of obtaining observed cell counts in matrix [A] that differ so much from the expected counts in matrix [E]. This low P-value gives strong evidence to reject H_0 in favor of the alternative that the proportions are not all the same.

The 2-PropZTest

Is it possible that the proportions of students who smoke are the same for the first two parent group types? We can test just these two types with a 2×2 table using the chi-square test, or we can use the TI-83's **2-PropZTest** to test if the proportions are the same versus the alternative that they are different.

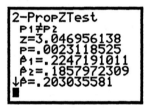

```
2-PropZTest              2-PropZTest
 x1:400                   P1≠P2
 n1:1780                  z=3.046956138
 x2:416                   P=.0023118525
 n2:2239                  p̂1=.2247191011
 P1:≠P2 <P2 >P2           p̂2=.1857972309
 Calculate Draw          ↓p̂=.203035581
```

The low P-value suggests that even just among these two parent group types, the proportions of students who smoke are different.

9.11 The Mediterranean diet. Cancer of the colon and rectum is less common in the Mediterranean region than in other Western countries. A 1953 Italian study compared the levels of consumption of olive oil among groups of patients to see if there was a possible difference in consumption levels for those who were cancer free. Here are some of the data.

Olive Oil Consumption

	Low	Medium	High	Total
Colon cancer	398	397	430	1225
Rectal cancer	250	241	237	728
Controls	1368	1377	1409	4154

Is high olive oil consumption more common among patients without cancer than in patients with colon or rectal cancer? Test to see if the levels of consumption are the same for each type of patient.

Solution: Press **MATRX**, scroll right to **EDIT**, and press **1**. Enter the data into a 3 × 3 matrix **[A]** (the totals are not necessary). Execute the **TWOWAY** program then view matrix **[D]**.

Now observe the third row of **[D]**. Around 33% of patients without colon or rectal cancer had low consumption of olive oil, around 33% have medium, and around 33% had high consumption.

```
MATRIX[D]  3 ×4
[ .3249    .3241    .351   -
[ .3434    .331     .3255  :
[ .3293    .3315    .3392  -
```

For patients with colon cancer, 35% had high consumption of olive oil; thus, it does not appear that high olive oil consumption is more common among patients without colon or rectal cancer.

We now test to see if the levels of consumption are the same for each type of patient.

Execute the X^2-**Test** feature using matrix **[A]** for the **Observed** and matrix **[E]** for the **Expected**. Then compare matrix **[A]** with matrix **[E]**.

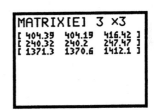

```
X²-Test
 X²=1.551580247
 P=.817467069
 df=4
```

```
MATRIX[A]  3 ×3
[ 398     397     430   ]
[ 250     241     237   ]
[ 1368    1377    1409  ]
```

```
MATRIX[E]  3 ×3
[ 404.35   404.19   416.42 ]
[ 240.32   240.2    247.47 ]
[ 1371.3   1370.6   1412.1 ]
```

We obtain a *P*-value of .817. If the levels of consumption are the same for each type of patient, then there would be over an 81% chance of obtaining observed cell counts in matrix **[A]** that differ by as much as these do from the expected counts. The observed differences are therefore due to random chance and are not statistically significant.

Thus, we can conclude that the proportion of patients having high olive oil consumption is the same for all three types of patients. We can conclude likewise for the proportion of those having medium consumption and for the proportion of those having low olive oil consumption.

Exercise 9.12 Preventing Strokes. During a two-year study, a comparison was made of different treatments for patients who had suffered a stroke.

Treatment	Number of patients	Number of strokes	Number of deaths
Placebo	1649	250	202
Aspirin	1649	206	182
Dipyridamole	1654	211	188
Both	1650	157	185

Make a two-way table of treatment by whether or not a patient had a stroke. Which treatment appears to be most effective in preventing strokes? Is there a significant difference among the four rates of strokes?

Solution: We first enter a two-way table in matrix [A]. The first column will represent the numbers of strokes, and the second column will represent the numbers of patients who did not have a stroke (1649 – 250, 1649 – 206, etc.).

After entering the data into matrix [A], execute the **TWOWAY** program and then observe matrix [D].

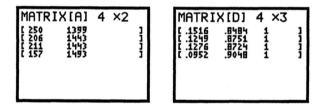

It appears that patients treated with both aspirin and dipyridamole had the lowest rate (.0952) of suffering a stroke.

We will now test to see if the proportions of patients having a stroke are the same regardless of treatment. Execute the X^2-**Test** feature using matrix [A] for the **Observed** and matrix [E] for the **Expected**. Then compare matrix [A] with matrix [E].

We obtain a *P*-value of 0.0000222. If the proportions of patients having a stroke were the same regardless of treatment, then there would be virtually no chance of obtaining the observed cell counts in matrix **[A]**. The observed differences are therefore statistically significant, and we must conclude that there is a significant difference among the four rates of strokes

We stress that it is still possible for a pair of treatments to have the same rate of stroke; it is just not the case that all four rates are the same. From matrix **[D]**, it appears that patients treated with just aspirin had the same rate of stroke as those treated with just dipyridamole. We can test this pair with the **2-PropZTest**.

Enter the data into the **2-PropZTest** feature, and observe the resulting *P*-value of .819. Clearly, we accept that the proportions of patients having a stroke are the same for these two treatments.

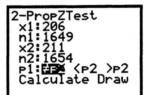

We can further test to see if treatment with both medicines results in a lower rate of stroke than treatment with just aspirin.

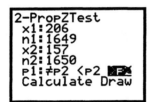
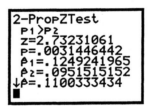

With the low *P*-value of .003, we can conclude that treatment with both medicines does result in a lower rate of stroke than treatment with just aspirin. If the rates were the same, then there would be only a .003 probability of the aspirin treatment stroke rate being so much higher than the dual treatment stroke rate.

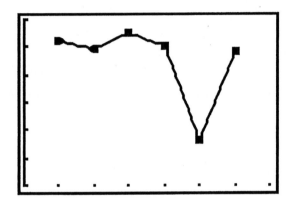

One-Way Analysis of Variance: Comparing Several Means

In this chapter we perform one-way analysis of variance (ANOVA) to test whether several normal populations, assumed to have the same variance, also have the same mean.

10.1   The Analysis of Variance *F* Test

We first work an example that uses the TI-83's built-in **ANOVA(** command (item **F)** from the **STAT TESTS** menu. This command will compute the *F*-statistic, the *P*-value, and the "pooled deviation" that can be used as an estimate of the populations' common standard deviation.

10.17 Nematodes and tomato plants. To see whether nematodes affect plant growth, 16 planting pots with tomato seedlings and various numbers of nematodes were studied for 16 days. The growth of each seedling (in cm) is listed below.

Nematodes	Seedling Growth			
0	10.8	9.1	13.5	9.2
1,000	11.1	11.1	8.2	11.3
5,000	5.4	4.6	7.4	5.0
10,000	5.8	5.3	3.2	7.5

(a) Compute the means and standard deviations of the four treatments and plot the means. (b) Perform an ANOVA test.

Solution: The **ANOVA(** command (item **F** in the **STAT TESTS** menu) requires the data to be in lists. Enter the data for each level of nematodes into lists **L1, L2, L3,** and **L4.**

(a) Since the treatments have the same sample size, we can compute the means and standard deviations pairwise with the **2-Var Stats** command from the **STAT CALC** menu.

Means and standard deviations for first pair of treatments

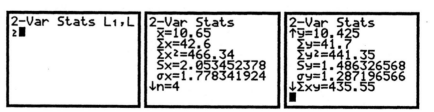

Means and standard deviations for third and fourth treatments

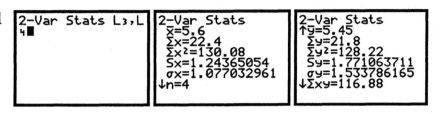

The values of the means are respectively 10.65, 10.425, 5.6, and 5.45. The sample deviations are respectively 2.0534, 1.4863, 1.2436, and 1.771.

To plot the means, we will make a time plot using lists **L5** and **L6**. Enter the integers 1 through 4 into **L5,** and enter the means into **L6**. Adjust the **WINDOW** and **STAT PLOT** settings, then graph.

(b) We will now test whether the true mean seedling growth is the same for each level of nematodes. The null hypothesis H_0 is that all the means are equal, and the alternative is that there is at least one pair of treatments that have different means.

Call up the **ANOVA(** command from the **STAT TESTS** menu, and enter the command **ANOVA(L1, L2, L3, L4)**.

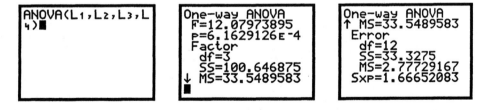

The F-statistic is given as 12.0797, which creates a P-value of 0.000616. If all the means were equal, then there would be only a 0.000616 probability of the sample means varying by as much as they do. We have significant evidence to reject the claim that all means are equal.

We can now say that there is at least one pair of treatments that have different means, although some pairs may have equal means. From the time plot, it appears that the first and second treatments may have the same mean and that the third and fourth treatments may have the same mean; however, the first and fourth clearly should have different means.

We can quickly test various pairs using the **2-SampTTest** feature. We shall now test the treatment pairs (**L1, L2**), (**L2, L3**).

Call up the **2-SampTTest** feature (item **4**) from the **STAT TESTS** menu. Set the lists to **L1** and **L2**, use a two-tail alternative, and pool the deviation, since we are assuming a common variance.

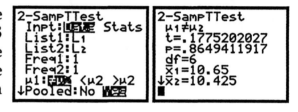

With the high *P*-value, we can conclude that there is no significant difference in the means from the first two treatments.

For the pair (**L2, L3**), we shall test for equality of mean, with the one-sided alternative that the mean from the second treatment is higher.

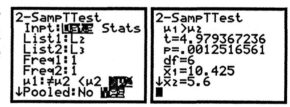

With a *P*-value of .00125, we see that the difference in means is statistically significant. If the means from these treatments were equal, there would be a very small chance of the sample mean from the second treatment being so much higher than the sample mean from the third treatment. So we reject that they are equal.

If desired, we can use the **2-SampTInt** feature to find confidence intervals for the differences in means between various pairs.

Finally note that in the original ANOVA test, a common pooled deviation for the four treatments is given as Sxp = 1.66652.

When the raw data are given, we can enter the data into lists and use the built-in **ANOVA(** command to test for equality of means. However, sometimes the summary statistics are given instead. The **ANOVA1** program computes the pooled deviation and the P-value of the ANOVA test in this case. Before executing the program, enter the sample sizes into **L1**, the sample means into **L2**, and the sample deviations into **L3**.

```
PROGRAM:ANOVA1
:1-Var Stats L₂,L₁
:sum(seq(L₁(I)(L₂(I)-x̄)²,I,1,dim(L₁))→A
:sum(seq((L₁(I)-1)(L₃(I)²),I,1,dim(L₁))→B
:dim(L₁)→I
:(A/(I-1))/(B/(n-I))→F
:1-Fcdf(0,F,I-1,n-I)→P
:√(sum(seq((L₁(I)-1)(L₃(I)²),I,1,dim(L₁))/(n-I)→S
:Disp "POOLED DEVIATION"
:Disp S
:Disp "P VALUE"
:Disp P
```

10.8 Who succeeds in college? The following table lists various high school statistics for several groups of students under study at a university.

Group	n	High school class rank $\bar{x}$	s	Semesters of HS math $\bar{x}$	s	Avg grade in HS math $\bar{x}$	s
CS majors	103	88.0	10.5	8.74	1.28	3.61	0.46
Sci./Eng. majors	31	89.2	10.8	8.65	1.31	3.62	0.40
Other	122	85.8	10.8	8.25	1.17	3.35	0.55

For each of the three high school variables, apply the ANOVA test to see if the representative larger populations have the same mean.

Solution: Enter the summary statistics for the class rank variable into lists **L1**, **L2**, and **L3**, and execute the **ANOVA1** program.

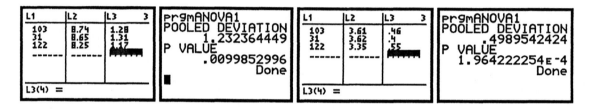

With a *P*-value of .1564, we see that the sample means are not significantly different. We do not have evidence to reject the claim that the average high school class ranks are the same for these three populations of students.

Next enter the summary statistics for the other variables and reexecute the **ANOVA1** program.

For both of these variables, the low *P*-values provide strong evidence that the means are not all equal. If the average number of semesters of high school math taken by these types of students were all the same, then there would be less than a 1% chance of the sample means varying by as much as they do. Likewise, we can conclude that the average high school math GPA is not the same for all students classified into these three types.

10.2 Some Details of ANOVA

We conclude with a final example that adds another detail in the event that we do not we reject the ANOVA hypothesis.

10.11 How much corn should I plant? Below are the sample means and sample deviations (from Figure 10.4) for the yield (in bushels per acre) from Exercise 10.2.

Plants per acre	n	$\bar{x}$	s
12,000	4	131.03	18.09
16,000	4	143.15	19.79
20,000	4	146.22	15.07
24,000	3	143.07	11.44
28,000	2	134.75	22.27

(a) Verify the result of the ANOVA test. Then calculate the overall mean yield. (b) Give a 90% confidence interval for the mean yield of corn planted at 20,000 plants per acre.

Solution: (a) We shall again use the **ANOVA1** program. First enter the summary statistics into lists **L1, L2,** and **L3,** then execute the program.

From the high *P*-value, we can accept the claim that the mean yield per acre is the same regardless of the number of plants per acre.

To find an estimate of the overall mean, we must weight the individual sample means. To do so, enter the command **1-VarStats L2, L1,** which will average the sample means in list **L2** that occur with the frequencies listed in **L1.**

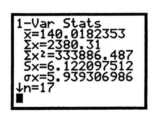

We see that an estimate of the overall mean is $\bar{x}$ = 140.018. This value can also be obtained by averaging all 17 measurements given in the raw data in Exercise 10.2.

(b) To find a 90% confidence interval for the mean yield of corn planted at 20,000 plants per acre, we use the **TInterval** feature from the **STAT TESTS** menu. We note that now we can used the pooled standard deviation value of 17.31014009 as an estimate of σ.

```
TInterval               TInterval
Inpt:Data MSTAT         (125.85,166.59)
x:146.22                x=146.22
Sx:17.31014009          Sx=17.31014009
n:4                     n=4
C-Level:.9
Calculate
```

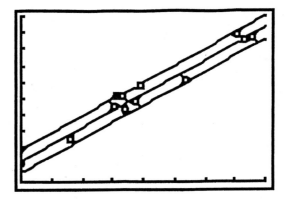

Inference for Regression

11.1 Inference about the Model
11.2 Inference about Prediction

In this chapter we provide details on using the TI-83 to perform the calculations for linear regression. In particular, we again find and graph the least-squares regression line and compute the correlation. We then test the hypothesis that the regression slope is equal to 0. We also provide a program that computes confidence intervals for the regression slope and intercept and another program that computes a prediction interval for a future observation and a confidence interval for a mean response.

11.1 Inference about the Model

We begin by using another command to find a regression line, while simultaneously testing the hypothesis that the regression slope equals 0.

11.2 Natural gas consumption. The following table gives the monthly gas consumption (in hundreds of cubic feet) of a household per average number of heating degree-days.

Month	Degree-days	Gas (100 cu ft)	Month	Degree-days	Gas (100 cu ft)
Nov.	24	6.3	July	0	1.2
Dec.	51	10.9	Aug.	1	1.2
Jan.	43	8.9	Sept.	6	2.1
Feb.	33	7.5	Oct.	12	3.1
Mar.	26	5.3	Nov.	30	6.4
Apr.	13	4.0	Dec.	32	7.2
May	4	1.7	Jan.	52	11.0
June	0	1.2	Feb.	30	6.9

(a) Make a scatterplot of the data. Find the correlation and the equation of the least-squares line. (b) Find the residuals and check that their sum is 0 (up to round-off error). (c) Estimate the three parameters α, β, and σ of regression inference.

Solution: Parts (a) and (b) are a review of the material covered in Sections 2.1 through 2.4. As before, we should make sure that the **DiagnosticOn** command has been entered from the **CATALOG**.

Now enter the data into lists. We will use **L1** for the degree-days x and **L2** for the gas consumption y. After entering the data, we choose an appropriate **WINDOW**, adjust the **STAT PLOT** settings, and **GRAPH**.

To find the correlation, press **STAT**, scroll right to **CALC**, press **8**, and then enter the command **LinReg(a+bx) L1, L2**. We see that $r \approx .995$, which depicts the strong linear relationship observed in the scatterplot.

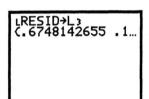

This command also gives us the equation of the least-squares line $y = a + bx \approx 1.089 + 0.189\,x$.

Note: Item **4** in the **STAT CALC** menu will give the least-squares line in the form $y = ax + b$. Here we have used item **8** to be consistent with the standard way of writing the regression line $\mu_y = \alpha + \beta x$.

(b) After computing a regression, the residuals are stored in the **LIST** screen. Press **2nd STAT** (**LIST**), and press **1** to obtain the command **LRESID** on the Home screen. Enter the command **LRESID →L3** to store the residuals into **L3**.

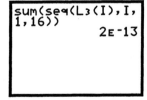

We can now verify that the residuals sum to 0. Press **2nd STAT**, scroll right to **MATH**, and press **5** to bring the command **sum(** to the Home screen. Next press **2nd STAT**, scroll right to **OPS**, and press **5** for the command **seq(** to make the command on the Home screen become **sum(seq(**. Now we wish to sum the 16 terms in list **L3**. Complete typing the command **sum(seq(L3(I), I, 1, 16))** and press **ENTER**.

```
NAMES OPS MATH
1:min(
2:max(
3:mean(
4:median(
5:sum(
6:prod(
7↓stdDev(
```

```
NAMES OPS MATH
1:SortA(
2:SortD(
3:dim(
4:Fill(
5:seq(
6:cumSum(
7↓ΔList(
```

```
sum(seq(L3(I),I,
1,16))
              2E⁻13
```

We see that the sum of the residuals is 0 up to round-off error.

(c) The regression parameters α and β are estimated by the regression coefficients a and b; thus, $\alpha \approx 1.089$ and $\beta \approx 0.189$. We can estimate σ directly by using the formula for s given by

$$s = \sqrt{\frac{1}{n-2}\sum residual^2}\,.$$

We can evaluate this sum as we summed the residuals in part (b):

```
√(sum(seq(L₃(I)²
,I,1,16))/14)
            .3389283987
```

We can also use the **LinRegTTest** feature (item E) from **STAT TESTS** menu as an alternate way to compute these regression coefficients.

```
LinRegTTest
 Xlist:L₁
 Ylist:L₂
 Freq:1
 β & ρ:≠0 <0 >0
 RegEQ:Y₁
 Calculate
```
```
LinRegTTest
 y=a+bx
 β≠0 and ρ≠0
 t=38.30862462
 P=1.415178ε-15
 df=14
↓a=1.089210843
■
```
```
LinRegTTest
 y=a+bx
 β≠0 and ρ≠0
↑b=.1889989538
 s=.3389283987
 r²=.9905504416
 r=.995264006
```

The coefficients a, b, s, and r are all given. This feature will be used in greater detail when we test the hypothesis of no linear relationship between the x and y measurements.

Confidence Intervals for Slope and Intercept

It is quite tedious to perform the calculations for the finding confidence intervals for the regression slope and intercept. Therefore we will use a program **REG1** to do this work for us.

Before executing the program, we must enter our data into lists and execute the **LinRegTTest** feature.

```
PROGRAM:REG1
:Disp "CONF. LEVEL"              :s√(1/n+x̄²/(nσx²))→A
:Input R                         :Disp "INT. FOR INT"
:"tcdf(0,X,n-2)"→Y₂              :Disp round({a-TA,a+TA},4)
:solve(Y₂-R/2,X,2)→T             :Disp "INT. FOR SLOPE"
:s/√(nσx²)→B                     :Disp round({b-TB,b+TB},4)
```

11.4 Natural gas consumption. Using the data of Exercise 11.2, find 95% confidence intervals for the slope and the intercept of the true regression line.

Solution: If not already done, execute the **LinRegTTest** feature from the **STAT TESTS** menu. Then call up the **REG1** program and enter .95 for **CONF. LEVEL**. The program takes a few seconds to execute while searching for the critical $t*$ values. Then the interval for the intercept and the interval for the slope are displayed.

In particular, a 95% confidence interval for the true regression slope is (.1784, .1996).

```
CONF. LEVEL
?.95
INT. FOR INT
   (.7913 1.3872)
INT. FOR SLOPE
   (.1784 .1996)
            Done
```

Testing the Hypothesis of No Linear Relationship

We shall now use the **LinRegTTest** feature to test the hypothesis H_0: $\beta = 0$, which is equivalent to testing whether there is no correlation between the variables x and y.

11.6 An extinct beast. The lengths of two bones in five fossil specimens of *Archaeopteryx* are given in the table.

Femur	38	56	59	64	74
Humerus	41	63	70	72	84

(a) Give the equation of the least-squares regression line. (b) Test the hypothesis H_0: $\beta = 0$ versus the alternative H_a: $\beta > 0$. Give the P-value of the test.

Solution: We can do the entire problem with the **LinRegTTest** feature (item E) in the **STAT TESTS** menu. First enter the data into lists, then adjust the list settings and alternative in the **LinRegTTest** feature and calculate.

```
L2     L3   L4    4    LinRegTTest          LinRegTTest          LinRegTTest
6.3    38   41         Xlist:L3             y=a+bx               y=a+bx
10.9   56   63         Ylist:L4             B>0 and p>0          B>0 and p>0
8.9    59   70         Freq:1               t=15.94050984        ↑b=1.196900115
7.5    64   72         B & p:≠0 <0 >0       P=2.6842019E-4       s=1.982027939
5.3    74   84         RegEQ:Y1             df=3                 r²=.9883313819
4      ------          Calculate            ↓a=-3.659586682      r=.9941485714
1.7
L4(6) =
```

The least-squares line is $y \approx -3.66 + 1.197x$. The P-value is given as .00026842, which is very low. If the true slope β were equal to 0, then there would be almost no chance of b being as large as 1.197.

Likewise, if the true correlation ρ were equal to 0, then there would be almost no chance of the sample correlation R being as large as .994. Thus we can reject the hypothesis in favor of the alternative. We are therefore saying that there is a linear relationship between the variables.

11.8 Does fast driving waste fuel? Following are the data set from Exercise 2.6 on the fuel used (liters/100 km) versus speed (km/h) of a small car. Is there evidence of straight-line dependence between speed and fuel use?

Speed	Fuel used	Speed	Fuel used	Speed	Fuel used
10	21.00	60	5.90	110	9.03
20	13.00	70	6.30	120	9.87
30	10.00	80	6.95	130	10.79
40	8.00	90	7.57	140	11.77
50	7.00	100	8.27	150	12.83

Solution: To analyze the data, we shall use lists **L5** and **L6**. We shall test the hypothesis H_0: $\beta = 0$ versus the alternative H_a: $\beta \neq 0$.

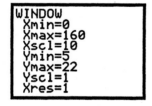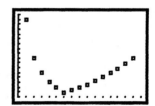

With the high P-value of .54, we do not have evidence to reject a claim that the regression slope β is equal to 0. If β were 0, there could still be 54% chance of the coefficient b being as far away from 0 as $-.014567$. Thus we do not have evidence of a straight-line relationship. A scatterplot verifies that there is no linear relationship.

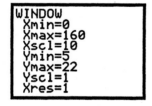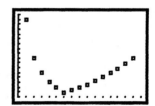

11.2 Inference about Prediction

We now provide another program **REG2** which computes a confidence interval for a mean response and a prediction interval for an estimated response.

Before this program can be executed, data must be entered into lists and the **LinRegTTest** must be performed on the calculator.

```
PROGRAM:REG2
:Disp "MEAN VALUE OF X?"        :s√(1/n+(X-x̄)²/(nσx²))→C
:Input X                        :s√(1+1/n+(X-x̄)²/(nσx²))→D
:Disp "FUT. VAL. OF X?"         :a+bX→Y
:Input W                        :a+bW→Z
:Disp "CONF. LEVEL"             :Disp "INT. FOR MEAN"
:Input R                        :Disp round({Y-TC,Y+TC},3)
:"tcdf(0,X,n-2)"→Y₂             :Disp "PREDICTION INT."
:solve(Y₂-R/2,X,2)→T            :Disp round({Z-TD,Z+TD},3)
```

11.10 Natural gas consumption.   Using the data of Exercise 11.2, give a 95% confidence interval for the predicted amount of gas that would be used for 40 degree-days per day. Give a 95% confidence interval for the mean gas consumption per day in all months with 40 degree-days per day.

Solution: We will use the data in lists **L1** and **L2**. Before executing the program, recalculate the **LinRegTTest** to compute the necessary statistical estimates. We can then compute both confidence intervals simultaneously with the **REG2** program. In this case both the future value for the prediction interval and the mean value for the mean response interval are $x^* = 40$. The resulting intervals are given below.

```
LinRegTest        LinRegTest         prgmREG2              CONF. LEVEL
Xlist:L1          y=a+bx             MEAN VALUE OF X?      ?.95
Ylist:L2          β≠0 and ρ≠0        ?40                   INT. FOR MEAN
Freq:1            t=38.30862462      FUT. VAL. OF X?          (8.388 8.91)
β & ρ:≠0 <0 >0    p=1.415178E-15     ?40                   PREDICTION INT.
RegEQ:Y1          df=14              CONF. LEVEL             (7.877 9.421)
Calculate         ↓a=1.089210843     ?.95                          Done
```

11.13 The endangered manatee. Here are data on powerboat registrations (in thousands) and the number of manatees killed by boats in Florida in the years 1977 to 1990.

Year	Powerboat registrations	Manatees killed	Year	Powerboat registrations	Manatees killed
1977	447	13	1984	559	34
1978	460	21	1985	585	33
1979	481	24	1986	614	33
1980	498	16	1987	645	39
1981	513	24	1988	675	43
1982	512	20	1989	711	50
1983	526	15	1990	719	47

Make a scatterplot showing the relationship between powerboat registrations (in thousands) and the number of manatees killed. Give a 90% confidence interval for the slope of the true regression line.

Solution: We will first enter the data into lists **L3** and **L4**. We then adjust the **WINDOW**, adjust the **STAT PLOT** settings, and **GRAPH**.

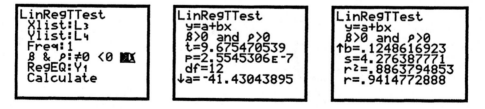

We then call up the **LinRegTTest** feature, adjust the settings, and calculate.

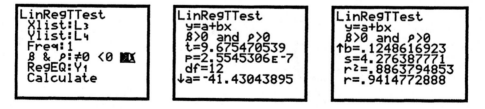

We note that the least-square line is $y = a + bx \approx -41.43 + 0.12486\,x$. Also note that for a null hypothesis $H_0: \beta = 0$ versus the alternative $H_a: \beta > 0$ the P-value is .0000002554. This low value gives strong evidence to reject H_0 and conclude that $\beta > 0$.

Since we have performed the **LinRegTTest**, we can now execute the **REG1** program. Call up the program and enter .90 for **CONF. LEVEL**. The interval for the slope is given as (.102, .148).

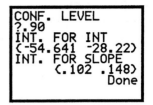

Note: Before executing the **REG1** program in this problem, we have edited the code from the **PRGM EDIT** screen so that the output displays are rounded to three decimal places.

11.14 Manatee predictions. With the data from Exercise 11.13, give a 95% confidence interval in a future year for which boat registrations are 716,000.

Solution: Again, since we have performed the **LinRegTTest** on the data in lists **L3** and **L4**, and assuming we have not yet computed other statistics with built-in TI-83 commands, we can execute the **REG2** program. (If other statistics have been computed, then quickly reexecute the **LinRegTTest**.)

Although we are asked only to find a prediction interval for $x^* = 716$, the **REG2** program will compute both the prediction interval and a mean response interval. We will enter 716 for both prompted variables.

The 95% prediction interval is given as (37.461, 58.48).

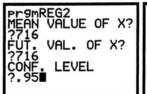

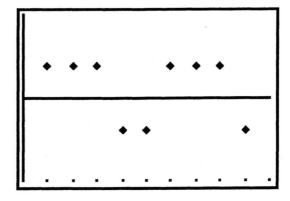

Nonparametric Tests

In this chapter we provide some supplementary programs for performing several nonparametric hypothesis tests.

12.1 The Wilcoxon Rank Sum Test

We first provide a program **RANKSUM** to perform the Wilcoxon rank sum test on data from two populations. To execute the program, we must enter the data into lists **L1** and **L2**. The program will first sort each list, then merge and sort the lists into list **L3**. Then it will put the rank of each measurement in **L3** next to it in **L4**. All sequences of ties are assigned an average rank. The sums of the ranks from **L1** and **L2** are next given.

The Wilcoxon test statistic W is the sum of the ranks from **L1**. Assuming that the two populations have the same continuous distribution (and no ties occur), then W has a mean and standard deviation given by

$$\mu = \frac{n_1(N+1)}{2} \quad \text{and} \quad \sigma = \sqrt{\frac{n_1 n_2 (N+1)}{12}},$$

where n_1 is the sample size from **L1**, n_2 is the sample size from **L2**, and $N = n_1 + n_2$.

We test the null hypothesis H_0: no difference in distributions. A one-sided alternative is H_a: first population yields higher measurements. We use this alternative if we expect or see that W is a much higher sum, which would also make W more than μ. In this case, the P-value is given by a normal approximation. We let $X \sim N(\mu, \sigma)$ and compute the right-tail $P(X \geq W)$ (using continuity correction if W is an integer).

If we expect or see that W is much lower than μ, then we should use the alternative H_a: first population yields lower measurements. In this case, the P-value is given by the left-tail $P(X \leq W)$, again using continuity correction if needed.

If the two sums of ranks are close, we could use a two-sided alternative H_a: there is a difference in distributions. In this case, the P-value is given by twice the smallest tail value ($2 P(X \geq W)$ if $W > \mu$, or $2 P(X \leq W)$ if $W < \mu$).

The **RANKSUM** program displays the smallest tail value created by the test statistic. It displays $P(X \geq W)$ if $W > \mu$, and it displays $P(X \leq W)$ if $W < \mu$. Conclusions for any alternative can then be drawn from this value.

We note that if there are ties, then the validity of this test is questionable.

```
PROGRAM:RANKSUM                  :Goto 1
:ClrList L₄                      :End
:SortA(L₁)                       :End
:SortA(L₂)                       :If I=M+N
:augment(L₁,L₂)→L₃               :Then
:SortA(L₃)                       :M+N→L₄(I)
:dim(L₁)→M                       :End
:dim(L₂)→N                       :1→I
:L₃→L₆                           :0→S
:L₃(1)-1→L₃(M+N+1)               :0→J
:1→B                             :Lbl 3
:1→I                             :While I≤M
:Lbl 1                           :Lbl 4
:While I<(M+N)                   :If L₁(I)=L₃(I+J)
:If L₃(I)<L₃(I+1)                :Then
:Then                            :S+L₄(I+J)→S
:B→L₄(I)                         :Else
:1+I→I                           :1+J→J
:1+B→B                           :Goto 4
:Goto 1                          :End
:Else                            :I+1→I
:1→J                             :Goto 3
:B→S                             :End
:Lbl 2                           :L₆→L₃
:While L₃(I)=L₃(I+J)             :ClrList L₆
:S+B+J→S                         :(M+N)(M+N+1)/2→R
:1+J→J                           :M*(M+N+1)/2→U
:Goto 2                          :√(M*N*(M+N+1)/12)→D
:End                             :(abs(S-U)-.5)/D→Z
:S/J→T                           :If int(S)≠S:(abs(S-U))/D→Z:End
:For(K,0,J-1)                    :.50-normalcdf(0,Z,0,1)→P
:T→L₄(I+K)                       :Disp "SUMS OF RANKS"
:End                             :Disp {S,R-S}
:I+J→I                           :If S=U:Disp "NO DIFFERENCE"
:B+J→B                           :If S<U:Disp "LEFT TAIL",round(P,4)
                                 :If S>U:Disp "RIGHT TAIL",round(P,4)
```

12.6 Does polyester decay? Below are breaking strengths (in pounds) of polyester strips buried for 2 weeks and of strips buried for 16 weeks.

2 weeks	118	126	126	120	129
16 weeks	124	98	110	140	110

(a) Apply the Wilcoxon rank sum test and compare the result with the P-value of 0.189 obtained from the **2-SampTTest**.
(b) What are the null and alternative hypotheses for the t test? For the Wilcoxon test?

Solution: We must enter the data into lists **L1** and **L2**. The Wilcoxon test statistic is the sum of ranks from **L1** in which we will enter the 2-week measurements. After entering the data, call up and execute the **RANKSUM** program. The result follows. The list **L3** now contains the merged and sorted measurements and **L4** contains the (averaged) ranks.

For the t test, we would use H_0: $\mu_1 = \mu_2$, H_a: $\mu_1 > \mu_2$, where μ_1 is the mean from all 2-week measurements and μ_2 is the mean from all 16-week measurements. For the Wilcoxon test, we use H_0: same distribution for both groups, H_a: 2-week measurements are higher.

The sum of the ranks of the 2-week measurements is 33, which is higher than the expected average of $\mu = 5(11)/2 = 27.5$. According to the Wilcoxon test, if the distributions were the same, then there would still be a .1481 probability (from the right-tail P-value) of the 2-week sum of ranks being so much higher than the expected average of 27.5. We therefore do not have significant evidence to reject H_0 in favor of the alternative.

In this case, the Wilcoxon P-value is slightly lower than the t test P-value; however, both are high enough to result in the same conclusion.

12.8 Logging in the rain forest. Below is a comparison of the number of tree species in unlogged plots in the rain forest of Borneo with the number of species in plots logged eight years earlier.

Unlogged	22	18	22	20	15	21	13	13	19	13	19	15
Logged	17	4	18	14	18	15	15	10	12			

Does logging significantly reduce the mean number of species in a plot after eight years? State the hypotheses, do a Wilcoxon rank sum test, and state the conclusion.

Solution: We will test the hypothesis H_0: no difference in mean versus the alternative H_a: unlogged mean is higher. To do so, we first enter the unlogged measurements into list **L1** and the logged measurements into list **L2**. We then execute the **RANKSUM** program and receive the following output.

We note that there are 21 measurements with 12 unlogged measurements. If there were no difference in mean, then we would expect the sum of ranks from **L1** to be $\mu = 12(22)/2 = 132$. If there were no change in mean, then there would be only a 2.98% chance of the sum of ranks from **L1** being as high as 159. This low P-value gives significant evidence to reject H_0 in favor of the alternative.

12.2 The Wilcoxon Signed Rank Test

We next provide the **SIGNRANK** program (page 140) to perform the Wilcoxon signed rank test on data sets of size n from two populations. To execute the program, we must enter the data into lists **L1** and **L2**. The program will sort the absolute value of the differences **L2 – L1** into list **L3**, but it will disregard any zero differences. The population size n is decreased to count only the nonzero differences.

```
PROGRAM:SIGNRANK
:ClrList L₃,L₄,L₅
:0→L₄(1):0→L₅(1):0→L₃(1)
:0→S:0→R:0→U:0→V:1→J
:For(I,1,dim(L₁),1)
:If L₂(I)-L₁(I)≠0
:Then
:abs(L₂(I)-L₁(I))→L₃(J)
:1+J→J
:End:End
:If L₃(1)>0
:Then
:SortA(L₃):dim(L₃)→N
:L₃→L₆:L₃(1)-1→L₃(N+1)
:1→B:1→I
:Lbl 1
:If I<N
:Then
:If L₃(I)<L₃(I+1)
:Then
:B→L₄(I)
:1+I→I:1+B→B
:Goto 1
:Else
:1→J:B→S
:Lbl 2
:If L₃(I)=L₃(I+J)
:Then
:S+B+J→S:1+J→J
:Goto 2
:End
:S/J→T
:For(K,0,J-1)
:T→L₄(I+K)
:End
:I+J→I:B+J→B
:End
:Goto 1
:End
```

```
:If I=N
:Then
:N→L₄(I)
:End
:0→J
:For(I,1,dim(L₁))
:If (L₂(I)-L₁(I))>0
:Then
:1+J→J:abs((L₂(I)-L₁(I))→L₅(J)
:End:End
:SortA(L₅)
:1→I:0→J
:If L₅(1)>0
:Then
:Lbl 3
:While I≤dim(L₅)
:Lbl 4
:If L₅(I)=L₃(I+J)
:Then
:U+L₄(I+J)→U
:Else
:1+J→J
:Goto 4
:End
:I+1→I
:Goto 3
:End:End
:L₆→L₃:ClrList L₆
:N(N+1)/2→R:N(N+1)/4→U
:√(N(N+1)(2N+1)/24)→D
:(abs(U-U)-.5)/D→Z
:If int(U)≠U:(abs(U-U))/D→Z:End
:.50-normalcdf(0,Z,0,1)→P
:Disp "SUMS OF SIGNED"
:Disp "RANKS -,+"
:Disp {R-U,U}
:If U=U:Disp "NO DIFFERENCE"
:If U<U:Disp "LEFT TAIL",round(P,4)
:If U>U:Disp "RIGHT TAIL",round(P,4)
```

The program puts the rank of each measurement in **L3** next to it in **L4**. All sequences of ties are assigned an average rank. The sums of the ranks of the positive differences and the negative differences will be displayed after the program runs.

The Wilcoxon test statistic W is the sum of the ranks from the positive differences. Assuming that the two populations have the same continuous distribution (and no ties occur), then W has a mean and standard deviation given by

$$\mu = \frac{n(n+1)}{4} \text{ and } \sigma = \sqrt{\frac{n(n+1)(2n+1)}{24}}.$$

We test the null hypothesis H_0: no difference in distributions. A one-sided alternative is H_a: second population yields higher measurements. We use this alternative if we expect or see that W is a much higher sum, which means that there were more positive differences in **L2 − L1**. In this case, the P-value is given by a normal approximation. We let $X \sim N(\mu, \sigma)$ and compute the right-tail $P(X \geq W)$ (using continuity correction if W is an integer).

If we expect or see that W is the much lower sum, then there were more negative differences. Now we should use the alternative H_a: second population yields lower measurements. In this case, the P-value is given by the left-tail $P(X \leq W)$, again using continuity correction if needed.

If the two sums of ranks are close, we could use a two-sided alternative H_a: there is a difference in distributions. In this case, the P-value is given by twice the smallest tail value ($2P(X \geq W)$ if $W > \mu$, or $2P(X \leq W)$ if $W < \mu$).

The **SIGNRANK** program displays the smallest tail value created by the test statistic. It displays $P(X \geq W)$ if $W > \mu$, and it displays $P(X \leq W)$ if $W < \mu$. Conclusions for any alternative can then be drawn from this value.

Again we note that if there are ties, then the validity of this test is questionable.

12.12 and 12.14 Stepping up your heart rate. Below are the hearts rates of five subjects before and after three minutes of exercise.   Each subject was tested at two exercise rates. Does exercise at the low rate significantly raise the heart rate? Apply the Wilcoxon signed rank test to reach a conclusion. Show the assignment of ranks in the calculation of the test statistic.

	Low Rate		High Rate	
	Resting	Final	Resting	Final
1	60	75	63	84
2	90	99	69	93
3	87	93	81	96
4	78	87	75	90
5	84	84	90	108

Solution: We will test the hypothesis H_0: median heart rates are the same before and after low-rate exercise versus the alternative H_a: final median heart rate is higher.   Enter the five resting heart rates into **L1** and the five final heart rates into **L2**.

The alternative means that there should be more positive differences, so the sum of the positive ranks should be higher. Therefore, the *P*-value comes from the right-tail probability created by the test statistic.   Execute the **SIGNRANK** program to obtain the results.

```
PrgmSIGNRANK
SUMS OF SIGNED
RANKS -,+
          (0 10)
RIGHT TAIL
             .0502
             Done
■
```

```
L3      L4      L5      5
6       1       6
9       2.5     9
9       2.5     9
15      4       15
------  ------  ------

L5(5) =
```

We see that the sum of the ranks of the positive differences is much higher than that of the negative differences. If the median heart rate were the same before and after exercising, then there would be only a .0502 chance of the sum of positive ranks being so much higher. We have rather significant evidence to conclude that the heart rates have gone up after low-rate exercise.

List **L3** now contains the ordered absolute values of the four nonzero differences. Their corresponding (averaged) ranks are adjacent in list **L4**. List **L5** contains only the positive differences. Since all the absolute differences were in fact positive differences, the test statistic is the sum of all the ranks in **L4**, which is 10.

12.20 Right versus left. Apply the Wilcoxon signed rank test to determine whether right-handed people can turn a right-threaded knob faster than they can a left-threaded knob. The times in seconds follow.

Subject	Right	Left	Subject	Right	Left
1	113	137	14	107	87
2	105	105	15	118	166
3	130	133	16	103	146
4	101	108	17	111	123
5	138	115	18	104	135
6	118	170	19	111	112
7	87	103	20	89	93
8	116	145	21	78	76
9	75	78	22	100	116
10	96	107	23	89	78
11	122	84	24	85	101
12	103	148	25	88	123
13	116	147			

Solution: We will test the null hypothesis H_0: no difference in times versus H_a: right-thread times are *faster*. We note that H_a is equivalent to saying that the left-thread times are *higher*. The tail probability for the P-value depends on which way we subtract the times. The **SIGNRANK** program always performs **L2 – L1**, but we can decide which group of measurements goes into which list.

If we enter the right-thread times into **L1** and the left-thread times into **L2**, then H_a is stating that **L2 – L1** causes more or higher-ranked positive differences. So the P-value comes from the right-tail probability. If we reverse the lists, then H_a is stating that there should be less or lower-ranked positive differences; thus the P-value comes from the left-tail probability.

To illustrate, we will enter the right-thread times into **L2** and the left-thread times into **L1**, then execute the **SIGNRANK** program.

Lists **L3** and **L4** show that there were 24 nonzero differences in time, and list **L5** shows that only five were positive differences, meaning that only five times were the right-thread times higher.

If there were no differences in time, then there would be only a .0038 chance of the sum of positive ranks being so low; thus, we can safely reject H_0 in favor of the alternative.

12.3 The Kruskal-Wallace Test

Our last program is for the Kruskal-Wallace test (page 145), which simultaneously compares the distribution of more than two populations. We test the null hypothesis H_0: all populations have same distribution versus the alternative H_a: measurements are systematically higher in some populations.

To apply the test, we draw independent SRSs of sizes $n_1, n_2, \ldots, n_i$ from I populations. There are N observations in all. We rank all N observations and let R_i be the sum of the ranks for the ith sample. The Kruskal-Wallace statistic is

$$H = \frac{12}{N(N+1)} \sum_{i=1}^{I} \frac{R_i^2}{n_i} - 3(N+1).$$

When the sample sizes are large and all I populations have the same continuous distribution, then H has an approximate chi-square distribution with $I-1$ degrees of freedom.

When H is large, creating a small right-tail probability P-value, then we reject the hypothesis that all populations have the same distribution.

12.22 and 12.24 Which color attracts beetles best? Use the Kruskal-Wallace test to see if there are significant differences in the numbers of insects trapped by the board colors.

Board Color	Insects trapped					
Lemon yellow	45	59	48	46	38	47
White	21	12	14	17	13	17
Green	37	32	15	25	39	41
Blue	16	11	20	21	14	7

```
PROGRAM:KRUSKAL                           :End:End
:ClrList L₃,L₄,L₅                         :If I=L
:dim([B])→L₁                              :Then
:sum(seq([B](1,J),J,1,L₁(2)))→L           :L→L₄(I)
:1→K                                      :End
:For(J,1,L₁(2))                           :1→K
:For(I,1,[B](1,J))                        :Lbl 5
:[A](I,J)→L₃(K)                           :While K≤L₁(2)
:1+K→K                                    :ClrList L₂
:End:End                                  :For(I,1,[B](1,K))
:SortA(L₃)                                :[A](I,K)→L₂(I)
;L₃→L₆:L₃(1)-1→L₃(L+1)                     :SortA(L₂)
:1→B:1→I                                  :End
:Lbl 1                                    :1→I:0→S:0→J
:While I<(L)                              :Lbl 3
:If L₃(I)<L₃(I+1)                          :While I≤[B](1,K)
:Then                                     :Lbl 4
:B→L₄(I)                                  :If L₂(I)=L₃(I+J)
:1+I→I:1+B→B                              :Then
:Goto 1                                   :S+L₄(I+J)→S:1+I→I
:Else                                     :Else
:1→J:B→S                                  :1+J→J
:Lbl 2                                    :End
:While L₃(I)=L₃(I+J)                       :Goto 3
:S+B+J→S:1+J→J                            :End
:Goto 2                                   :S→L₅(K):1+K→K
:End                                      :Goto 5
:S/J→T                                    :End
:For(K,0,J-1)                             :L₆→L₃:ClrList L₂,L₆
:T→L₄(I+K)                                :12/L/(L+1)*sum(seq(L₅(I)²/[B](1,I),
:End                                         I,1,L₁(2)))-3(L+1)→W
:I+J→I:B+J→B                              :1-x²cdf(0,W,L₁(2)-1)→P
:Goto 1                                   :Disp "TEST STAT",round(W,5)
                                          :Disp "P VALUE",round(P,5)
```

Solution: To execute the **KRUSKAL** program, we must use the **MATRX EDIT** screen to enter the data into matrix **[A]** and the sample sizes into matrix **[B]**. First enter the data into the columns of the 6 × 4 matrix **[A]** as you would normally enter data into lists. Then enter the sample sizes into a 1 × 4 matrix **[B]**.

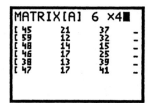

Next call up and execute the **KRUSKAL** program. Then view the entries in lists **L3**, **L4**, and **L5**. List **L3** contains the merged, sorted measurements, and list **L4** contains their (averaged) ranks. List **L5** contains the sum of ranks from each type of color.

The low *P*-value of .00072 is evidence to reject that all colors yield the same distribution of insects trapped.

INDEX OF PROGRAMS

ANOVA1 (page 121) Displays the pooled deviation and the P-value of the ANOVA test for equality of means when the data are entered as summary statistics. Before executing the program, enter the sample sizes into **L1**, the sample means into **L2**, and the sample deviations into **L3**.

BAYES (page 67) Computes the total probability $P(A)$ and the conditional probabilities associated with Bayes' rule. Before executing the program, enter the values of $P(B_i)$ into list **L1** and the conditionals $P(A|B_i)$ into list **L2**. The program displays $P(A)$, stores $P(A \cap B_i)$ in list **L3**, stores $P(B_i|A)$ in list **L4**, stores $P(B_i|A')$ in list **L5**, and stores $P(A|B_i')$ in list **L6**.

BINOMIAL (page 63) Computes the probability of a binomial distribution upon entering values for n and p, and for the lower and upper bounds. The program displays the probability along with the average value. Has an option to store the entire distribution in lists **L1**, **L2**, and **L3**.

GEOMET (page 52) Computes the probability of a geometric distribution upon entering values for p and for the lower and upper bounds. The program displays the probability along with the median and the average value. Has an option to store the entire distribution in lists **L1**, **L2**, and **L3**.

HYPGEOM (page 59) Computes the probability of a hypergeometric distribution upon entering values for the population size, type A size, sample size, and for the lower and upper bounds. The program displays the probability along with the average value and standard deviation. Has an option to store the entire distribution in lists **L1**, **L2**, and **L3**.

KRUSKAL (page 145) Performs the Kruskal-Wallace test. Before executing, enter the data into the columns of matrix **[A]** and the sample sizes into a row matrix **[B]**. The program displays the test statistic and P-value. Then **L3** contains the merged, sorted measurements, **L4** contains their (averaged) ranks, and **L5** contains the sum of ranks from each population.

NORMDIST (page 15) Displays the probability of a normal distribution upon entering values for the mean, standard deviation, lower bound, and upper bound. Displays the left-tail and right-tail values and the body area when entering the same value for the lower bound and the upper bound.

PSAMPSZE (page 100) When finding a confidence interval for a proportion, computes the required sample size that would provide a certain maximum margin of error m with a certain level of confidence.

RANKSUM (page 137) Performs the Wilcoxon rank sum test on data from two populations. Before executing, enter the data into lists **L1** and **L2**. The program displays the sums of the ranks from each list and the smallest tail value created by the test statistic that is the sum of the ranks from **L1**. List **L3** then contains the merged, sorted measurements, and **L4** contains their (averaged) ranks.

REG1 (page 128) Finds confidence intervals for the regression slope and intercept. Before executing the program, data must be entered into lists and the **LinRegTTest** (from the **STAT TESTS** menu) must be performed.

REG2 (page 131) Finds a confidence interval for a mean response and a prediction interval for an estimated response when performing linear regression. Before executing the program, data must be entered into lists and the **LinRegTTest** (from the **STAT TESTS** menu) must be performed.

SAMPLEN (page 38) Generates a random sample from a specified normal distribution and stores the data in list **L1**. Then displays the sample mean and sample deviation to compare with the true parameters.

SAMPLEN2 (page 43) Generates a specified number of random samples, each of the same specified sample size, from a specified normal distribution. Also computes the sample mean for each sample and stores it in list **L2**. Then displays the average and standard deviation of the sample means to compare with theoretical mean and standard deviation of the sampling statistics.

SAMPLEP (page 38) Generates count data for a specified proportion p and sample size n and stores the data in list **L1**. Then displays the sample proportion to compare with the true proportion.

SAMPLEP2 (page 42) Generates a specified number of random samples of count data, each of the same specified sample size and for the same proportion p. Also computes the sample proportion for each sample and stores it in list **L2**. Then displays the average of the sample proportions to compare with the real proportion p.

SIGNRANK (page 140) Performs the Wilcoxon signed rank test on data sets of size n from two populations. Before executing, enter the data into lists **L1** and **L2**. The program sorts the absolute value of the differences **L2 − L1** into list **L3**, but disregards any zero differences. The (averaged) rank of each non-zero difference is stored in **L4**. The sums of the ranks of the positive differences and of the negative differences are displayed. The program also displays the smallest tail value created by the test statistic, which is the sum of the ranks of the positive differences.

TSCORE (page 86)   Finds the critical value (t score) of a t distribution upon specifying the degrees of freedom and confidence level.

TWOWAY (page 33)   Converts a two-way table of raw data into three different proportion tables.   Before executing the program, enter the raw data into matrix **[A]**.   The proportion tables are stored in matrices **[B]**, **[C]**, and **[D]**.

ZSAMPSZE (page 76)   When finding a confidence interval for the mean using the normal distribution, computes the sample size needed to obtain a desired margin of error with a specified level of confidence.